D0436139

A Many-Colored Glass

Page-Barbour Lectures for 2004

FREEMAN J. DYSON

A Many-Colored Glass

Reflections on the Place of Life in the Universe

University of Virginia Press
Charlottesville and London

University of Virginia Press

Printed in the United States of America on acid-free paper
First published 2007

9 8 7 6 5 4 3 2 1

LIBRARY OF CONGRESS CATALOGING-IN-PUBLICATION DATA

Dyson, Freeman J.

A many-colored glass : reflections on the place of life in the universe / Freeman J. Dyson.

p. cm. — (Page-Barbour lectures for 2004)

Includes bibliographical references and index.

ISBN 978-0-8139-2663-6 (cloth : alk. paper)

1. Life (Biology)—Philosophy. 2. Biology—Philosophy. 3. Science—Philosophy. 4. Society—Philosophy. I. Title.

QH501.D97 2007

570.1—dc22

2006103104

Life, like a dome of many-colored glass,
Stains the white radiance of eternity.
—Percy Bysshe Shelley, "Adonais"

Contents

Preface

This book is a collection of lectures on the general theme of life in the universe. It began with three lectures given at the University of Virginia in March 2004. These were called Page-Barbour Lectures in honor of Thomas Page and the Barbour family of Virginia. The lectures were originally funded in 1907 by Thomas Page's wife, who belonged to the Barbour family. Thomas Page was a diplomat and writer, and most of the previous lecturers have been philosophers or historians or poets. Among my distinguished predecessors were T. S. Eliot and W. H. Auden. I am the first physical scientist to give the lectures. The text of my Page-Barbour lectures is to be found in chapters 2, 4, 6 and parts of chapter 1.

Since the three Page-Barbour lectures would have made a very skimpy book, I have supplemented them with other lectures given at various times and places. Chapter 3 and parts of Chapter 1 are taken from the Frederick S. Pardee Distinguished Lecture for fall 2005, given at the Frederick S. Pardee Center for the Study of the Longer-Range Future at Boston University in October 2005. I am grateful to the Pardee Center for permission to print the lecture here. Chapter 5 is from a lecture given at the Rockefeller University in June 2001. Chapter 7 is from a lecture given at the Center of

Theological Inquiry in Princeton in November 2003. I have edited these lectures so as to fit them together with the Page-Barbour lectures without overlapping. The result is a collection of essays rather than a coherent narrative.

The book has three main themes. The first theme is political, trying to understand the human and ethical consequences of biotechnology. The second theme is scientific, trying to understand intellectually the place of life in the universe. The third theme is personal, trying to understand the implications of biology for philosophy and religion. The first theme occupies chapters 1 to 3; the second, chapters 4 to 6; and the third, chapter 7. I have not tried to squeeze my thoughts about various aspects of science and technology into a unified theory. My reflections spill over from astronomy into ecology, and in the last chapter from neurology into theology. Life is, as Shelley said, a dome of many-colored glass, and its beauty lies in its variety.

I am grateful to the University of Virginia for the hospitality that I enjoyed while giving the Page-Barbour Lectures, and for the opportunity to publish them in this volume. I am also grateful to the institutions that gave me help and encouragement to compose the other lectures that are here included. I owe an apology to Amy Lowell, who published her first book of poems in 1912 with the title *A Dome of Many-Colored Glass.* I pray her forgiveness for taking a piece of her title. We both independently borrowed it from Shelley.

The three Page-Barbour lectures were given with the series title "Life in the Universe." The individual lectures were "A Friendly Universe" (chap. 4), "Looking for Life" (chap. 6), and "Possible Futures" (chaps. 1 and 2).

Chapter 3 and parts of chapter 1 were published as *Heretical Thoughts about Science and Society,* the Frederick S. Pardee Distinguished Lecture, November 1, 2005. This is part of the Distinguished Lecture Series published by the Frederick S. Pardee Center for the Study of the Longer-Range Future, at Boston University (© 2006 Boston University).

A more technical version of chapter 5 with the title "Is Life Digital or Analog?" was published in 2002 by the Templeton Foundation Press as a chapter in a book, *The Far-Future Universe.*

A more technical version of chapter 6 was published with the title "Looking for Life in Unlikely Places: Reasons Why Planets May Not Be the Best Places to Look for Life," in the *International Journal of Astrobiology* 2 (2003): 103–10.

Chapter 7 was published by the Center of Theological Inquiry in its magazine *Reflections* 7 (Spring 2004): 32–57. I learned after this book was finished that Octavia Butler died on February 24, 2006, at the age of fifty-eight. She fell dead on the path in front of her home, probably as the result of a stroke.

A Many-Colored Glass

1 The Future of Biotechnology

Hedgehogs and Foxes

Scientists come in two varieties, hedgehogs and foxes. I borrow this terminology from Isaiah Berlin (1953), who borrowed it from the ancient Greek poet Archilochus. Archilochus told us that foxes know many tricks, hedgehogs only one. Foxes are broad, hedgehogs are deep. Foxes are interested in everything and move easily from one problem to another. Hedgehogs are only interested in a few problems that they consider fundamental, and stick with the same problems for years or decades. Most of the great discoveries are made by hedgehogs, most of the little discoveries by foxes. Science needs both hedgehogs and foxes for its healthy growth, hedgehogs to dig deep into the nature of things, foxes to explore the complicated details of our marvelous universe. Albert Einstein and Edwin Hubble were hedgehogs. Charley Townes, who invented the laser, and Enrico Fermi, who built the first nuclear reactor in Chicago, were foxes. It often happens that foxes are as creative as hedgehogs. The laser was a big discovery made by a fox. The general public is misled by the media into believing that great scientists are all hedgehogs. Some periods in the history of science are good times for hedgehogs, other periods are good times for foxes. The beginning of the twentieth century was good for hedgehogs. The hedgehogs—Einstein and his followers in Europe, Hubble and his followers in America—dug deep and found new foundations for physics and astronomy. When Fermi and Townes came onto the scene in the middle of the century, the foundations were firm and the universe was wide open for foxes to explore. Most of the progress in physics and astronomy since the 1920s was made by foxes.

Another fox who played an important part in twentieth-century science was John von Neumann. Von Neumann was interested in almost everything and made important contributions to many fields. In the 1920s he found the first axiomatic formulation of set theory that was free of logical contradictions, an achievement that enabled his hedgehog friend Kurt Gödel to prove his famous theorem about the existence of undecidable propositions in arithmetic. Gödel continued to be a hedgehog and von Neumann continued to be a fox. After straightening out set theory, von Neumann invented game theory, found the first mathematically rigorous formulation of quantum mechanics, and studied the logical architecture of automatic machinery and human brains. Fifty years ago in Princeton, I watched him designing and building the first electronic computer that operated with instructions coded into the machine. He did not invent the electronic computer. The computer called ENIAC had been running at the University of Pennsylvania five years earlier. What von Neumann invented was software, the coded instructions that gave the computer agility and flexibility. It was the combination of electronic hardware with punch-card software that allowed a single machine to predict weather, to simulate the evolution of populations of living creatures, and to test the feasibility of hydrogen bombs.

I was lucky to be at the Institute for Advanced Study in Princeton in the 1940s and '50s when von Neumann's computer project began. He invited lively young people from all over the world to start new fields of science that the computer would make possible. The biggest group were meteorologists, who started the science of climate modeling. Another group were mathematicians, who started what later became known as computer science. Another group were hydrogen bomb designers, who brought their codes secretly to run on the machine during the midnight shift. And there was Nils Barricelli, a lone biologist who ran codes simulating biological evolution. He started the science of artificial life forty years before it became fashionable. Von Neumann was interested in all these activities, but most of all in meteorology. He had grand ideas about meteorology. I remember him giving a talk about the future of meteorology. He said,

> As soon as we are able to simulate the fluid dynamics of the atmosphere on a computer with adequate precision, we will be able to apply simple tests to decide whether the situation is stable or unstable. If the situation is stable, we can predict what will happen next. If the situation is unstable, we can apply a small perturbation to control what will happen next. The necessary perturbations can be applied by high-flying airplanes with smoke generators, warming the atmosphere where the smoke absorbs sunlight, and cooling the atmosphere in the shaded region underneath. So we shall be masters of the weather. Whatever we cannot control, we shall predict, and whatever we cannot predict, we shall control.

He estimated that it would take about ten years to develop a computer that would give us this kind of control over the weather.

Von Neumann, of course, was wrong. He was a great mathematician but a very poor predictor of the future. I have no illusion that I am a better predictor than he was. That is why in this chapter I am writing mostly about the past. The past is much easier to predict. Von Neumann was wrong because he did not know about chaos. He imagined that if a situation in the atmosphere was unstable, he could always apply a small perturbation to move it into a situation that was stable and therefore predictable. In fact this is not true. Most of the time, when the atmosphere is unstable, the motion is chaotic, which means that any small perturbation will only move it into another unstable situation which is equally unpredictable. When the motion is chaotic, it can be neither predicted nor controlled. So von Neumann's dream was an illusion. But the fact that the equations of meteorology have chaotic solutions was only discovered by the meteorologist Edward Lorenz at MIT in 1961, four years after von Neumann died.

Von Neumann made another prediction that also turned out to be wrong. He predicted, rightly, that his invention of electronic computers with programmable software would change the world. He understood that the descendants of his machine would dominate the operations of science and business and government. But he imagined computers growing larger and more expensive as they grew more powerful. He imagined them as giant centralized facilities serving large research laboratories or large industries. According to legend, somebody in the government once asked him how many computers the United States would need in the future, and he replied, "Eighteen." I do not know whether this legend has any foundation in fact. But it is certainly true that von Neumann had no inkling of the real future of computers. It never entered his head that computers would grow smaller and cheaper as they became faster and smarter. He never imagined computers becoming small enough and cheap enough to be used by housewives for doing income-tax returns and by children for doing homework. He failed totally to foresee the final domestication of computers as toys for three-year-olds. He failed to foresee the emergence of computer games as a dominant feature of twenty-first-century life. Because of computer games, our grandchildren are now growing up with an indelible addiction to computers. For better or for worse, in sickness or in health, till death do us part, humans and computers are now joined together more durably than husbands and wives.

Besides providing entertainment for our grandchildren, the domestication of computers has also provided the tools that make many small scientific enterprises possible. Cheap small computers have made it possible for small enterprises to make serious contributions to science and to compete successfully with big enterprises. Astronomers at a small observatory can discover an earthlike planet thousands of light-years away by measuring precisely the gravitational focusing by the planet of the light from a more distant star. Chemists working with apparatus on a tabletop can measure precisely the production of natural gas two hundred kilometers down in the mantle of the earth. Von Neumann's original computer in Princeton had a total memory capacity of four kilobytes. Nowadays, a scientist running a small project can easily afford a database of four gigabytes, a million times larger than von Neumann's memory and much cheaper. Big projects today have databases containing millions of gigabytes. Million-gigabyte memories are expensive and need staffs of experts to organize them efficiently. A little project that requires only a few gigabytes may have a competitive advantage. Foxes who organize small projects in their spare time can move ahead more rapidly than hedgehogs who devote their whole lives to big projects.

I am not predicting that the twenty-first century will be a golden age of foxes without any need for hedgehogs. I am saying that the history of science shows an alternation between times when hedgehogs are dominant and times when foxes are dominant. Hedgehogs were dominant in the seventeenth century, the age of Kepler and Newton. Foxes were dominant in the eighteenth century, the age of Euler and Franklin. Hedgehogs were dominant in the early twentieth century, the age of Einstein and Dirac. Foxes were dominant in the middle twentieth century, the age of Fermi and Townes. Maybe we are due now for another age of hedgehogs to shake up the foundations of science. Or maybe not. The future is unpredictable. In either case, whether or not hedgehogs return to cause a major scientific revolution, there will always be a need for foxes to carry on the normal business of science. In the coming century, no matter what the hedgehogs may be doing, the domestication of high technology will be giving new opportunities for foxes to achieve great results with limited means.

The Domestication of High Technology

Not only computers but also other scientific instruments of high precision have been domesticated during the last twenty years. The most spectacular case of domesticated high technology is the GPS or Global Positioning System. Twenty years ago, the GPS was a secret military program with location data available to civilians only in degraded form. Now the data with full accuracy are available to everybody, providing accurate location of the receiver in space and time at a price that ordinary hikers and sailors can afford. Likewise, digital cameras providing instant images of high quality are now for sale in every camera shop and are rapidly making film cameras obsolete. Digital cameras have also caused a revolution in astronomy. At first, when digital cameras were still experimental and expensive, they were used only at large professional observatories. But now, since digital cameras have been domesticated, they are used routinely at small observatories and by amateur astronomers. Digital cameras, combined with data processing by personal computers, allow amateurs and students to make precise scientific observations of a kind that could formerly be done only by professional astronomers with large instruments. The domestication of high technology will make small projects more and more cost-effective as time goes on.

It has become part of the accepted wisdom to say that the twentieth century was the century of physics and the twenty-first century will be the century of biology. Two facts about the coming century are agreed on by almost everyone. Biology is now bigger than physics, as measured by the size of budgets, by the size of the workforce, or by the output of major discoveries, and biology is likely to remain the biggest part of science through the twenty-first century. Biology is also more important than physics, as measured by its economic consequences, by its ethical implications, or by its effects on human welfare. These facts raise an interesting question. Will the domestication of high technology, which we have seen marching from triumph to triumph with the advent of personal computers and GPS receivers and digital cameras, soon be extended from physical technology to biotechnology? I believe that the answer to this question is yes. Here I am bold enough to make a definite prediction. I predict that the domestication of biotechnology will dominate our lives during the next fifty years at least as much as the domestication of computers has dominated our lives during the previous fifty years.

I see a close analogy between von Neumann's blinkered vision of computers as large centralized facilities and the public perception of genetic engineering today as an activity of large pharmaceutical and agribusiness corporations such as Monsanto. The public distrusts Monsanto because Monsanto likes to put genes for poisonous pesticides into food crops, just as we distrusted von Neumann because von Neumann liked to use his computer for designing hydrogen bombs secretly at midnight. It is likely that genetic engineering will remain unpopular and controversial so long as it remains a centralized activity in the hands of large corporations.

I see a bright future for the biotechnology industry when it follows the path of the computer industry, the path that von Neumann failed to foresee, becoming small and domesticated rather than big and centralized. The first step in this direction was already taken recently, when genetically modified tropical fish with new and brilliant colors appeared in pet stores. For biotechnology to become domesticated, the next step is to become user-friendly. I recently spent a happy day at the Philadelphia Flower Show, the biggest flower show in the world, where flower breeders from all over the world show off the results of their efforts. I have also visited the Reptile Show in San Diego, an equally impressive show displaying the work of another set of breeders. Philadelphia excels in orchids and roses, San Diego excels in lizards and snakes. The main problem for a grandparent visiting the reptile show with a grandchild is to get the grandchild out of the building without actually buying a snake. Every orchid or rose or lizard or snake is the work of a dedicated and skilled breeder. There are thousands of people, amateurs and professionals, who devote their lives to this business. Now imagine what will happen when the tools of genetic engineering become accessible to these people. There will be do-it-yourself kits for gardeners who will use genetic engineering to breed new varieties of roses and orchids. Also kits for lovers of pigeons and parrots and lizards and snakes to breed new varieties of pets. Breeders of dogs and cats will have their kits too.

Domesticated biotechnology, once it gets into the hands of housewives and children, will give us an explosion of diversity of new living creatures, rather than the monoculture crops that the big corporations prefer. New lineages will proliferate to replace those that monoculture farming and deforestation have destroyed. Designing genomes will be a personal thing, a new art form as creative as painting or sculpture. Few of the new creations will be masterpieces, but all will bring joy to their creators and variety to our fauna and flora. The final step in the domestication of biotechnology will be biotech games, designed like computer games for children down to kindergarten age but played with real eggs and seeds rather than with images on a screen. Playing such games, kids will acquire an intimate feeling for the organisms that they are growing. The winner could be the kid whose seed grows the prickliest cactus, or the kid whose egg hatches the cutest dinosaur. These games will be messy and possibly dangerous. Rules and regulations will be needed to make sure that our kids do not endanger themselves and others. The dangers of biotechnology are real and serious. I will discuss the dangers and possible remedies in the following chapter.

If domestication of biotechnology is the wave of the future, five important questions need to be answered. First, can it be stopped? Second, ought it to be stopped? Third, if stopping it is either impossible or undesirable, what are the appropriate limits that our society must impose on it? Fourth, how should the limits be decided? Fifth, how should the limits be enforced, nationally and internationally? I do not attempt to answer these questions here. I leave it to our children and grandchildren to supply the answers.

The actual shape of domesticated biotechnology is as impossible for us to discern today as the actual shape of a personal computer was impossible for von Neumann to discern in 1950. The best that I can do is to describe the functions of a do-it-yourself biotechnology kit. I cannot guess the shapes of the machines that will carry out the functions. The kit will have five chief functions. First, to grow plants under controlled conditions. This requires a garden or greenhouse with the usual tools and chemical supplies. Second, to grow animals under controlled conditions. This requires a stable for big animals or cages for small animals, with the usual supplies of food and medicaments. Third, simple and user-friendly instruments allowing unskilled people to manipulate seeds or eggs or embryos. Fourth, a tabletop genome sequencer able to sequence single molecules of DNA. DNA is the nucleic acid molecule that carries genetic information, and sequencing of the DNA in any creature means reading its genome. Fifth, a tabletop genome synthesizer able to synthesize substantial quantities of DNA with any desired sequence. The latter two instruments do not now exist, but they are likely to exist within ten or twenty years, since they will have great commercial value for pharmaceutical industries, and great practical value for medicine and scientific research.

What use will scientists make of these domesticated biotechnology kits when they become widespread? A good answer to this question was given by Herbert Kroemer of Santa Barbara, who won a Nobel Prize in the year 2000 for his invention of new kinds of semiconductor materials. He said in his Nobel lecture, "The principal applications of any sufficiently new and innovative technology always have been, and will continue to be, applications created by that technology" (in Cahn 2005). A great example illustrating the truth of this remark is the invention of the laser by Townes. Almost nothing that lasers now do was foreseen before they were invented. The applications of domesticated biotechnology will be at least as novel and diverse as the applications of laser technology. Domesticated biotechnology will begin with gardens and pets but will rapidly spread to infiltrate the operations of mines and factories, laboratories and supermarkets. Domesticated biotechnology will allow many objects of commerce and daily life, such as chairs and tables and houses and roads, to be grown rather than manufactured. When teenagers become as fluent in the language of genomes as they are fluent today in the language of blogs, they will be designing and growing all kinds of useful and useless works of art for fun and profit.

I do not venture to predict what new scientific revolutions will emerge from a mastery of biotechnology. One of the nightmares that I can imagine is that medical researchers will find a cure for death. After that, aged immortals will accumulate on this planet and there will be no room for the young. The normal replacement of each generation by the next will come to an end, and progress in science will stop. This is one way in which technology might put an end to science. A more hopeful outcome of biotechnology is the design and breeding of radically new microbes and plants and animals adapted to living wild in cold places such as Mars and the satellites of Jupiter and Saturn. New ecologies adapted to low levels of sunlight could bring these alien worlds to life. Plants that grow their own greenhouses could generate breathable air and keep the surfaces of these worlds warm, so that they would become hospitable to human settlement. In this way, new generations of young scientists could keep science alive in remote places while preserving planet Earth as a retirement home for aged immortals.

Open-Source Software

Fortunately, young scientists are still flourishing here on planet Earth. I met a number of them recently at a meeting in Portland, Oregon. This informal and enjoyable meeting was called OSCON, short for Open Source Convention. It was a meeting organized by a group of people who call themselves the Geek Culture. Many of them are people who dropped out of college and started software companies. There were about a thousand geeks at the meeting, mostly young and adventurous and interested in other things besides getting rich. One of the people I got to know at the meeting was Brewster Kahle, the founder of an enterprise called Internet Archive. His aim is to put the literature of the world in all languages into digital memory and make it accessible to everybody. He has made a good start with three databases, one in San Francisco, one in Amsterdam, and one in Alexandria on the site of the ancient library. He intends to have two more databases in India and China. Each of his databases will hold a copy of the entire archive, so that the heritage of world literature will survive even if four of the five databases are poorly maintained or destroyed. The archive will be a hundred million books, or a few million gigabytes of data. With modern memory storage, the archive can be housed in a room of modest size and costs far less than a big conventional library.

Kahle and the OSCON crowd share a belief in open-source software. That means that their companies are based on software programs that are out in the open like UNIX and LINUX, free for anyone to copy and improve. They share an intense dislike for companies like Microsoft that keep their source code secret. They observe that Bill Gates's proprietary software is full of bugs. Their open-source software has fewer bugs because all users are allowed and encouraged to debug it. The philosophy of Open Source is based on sharing. The way to achieve bug-free and user-friendly software is to share intellectual property freely. Within the Open Source community, people can build on each others' successes and avoid each others' mistakes.

I am not an expert on open-source software. I spoke in Portland about the linkage that I see between the Open Source movement in the software world and the Open Genome movement in the biological world. Just as the software industry is split between Open Source on the one side and Bill Gates on the other, the biotech world is split between the international research community believing in open databases on the one side and Craig Venter and William Hazeltine on the other. Venter and Hazeltine founded companies, Celera Genomics and Human Genome Sciences, respectively, that make money by selling access to genomic databases that they privately own. Both companies have made big investments in research and have published a large fraction of their discoveries. But they remain fundamentally committed to private ownership of intellectual property. Like Bill Gates, Venter and Hazeltine are philanthropists. Also like Bill Gates, they are philanthropists on their own terms. They do not give away the geese that lay the golden eggs.

I talked to the Open Source crowd about another kind of biological sharing. In addition to sharing genome databases, biological communities can also share genes. The physical sharing of genes between diverse members of a community gives another meaning to the phrase Open Source. When genes are shared freely, a biological community reaps the same advantages from sharing genes that the Open Source community reaps from sharing software. I was suggesting that the Open Source movement may be recapitulating in a few decades the history of life on Earth over billions of years.

A New Biology for a New Century

Carl Woese is the world's greatest expert in the field of microbial taxonomy, the classification and understanding of microbes. He explored the ancestry of microbes by tracing the similarities and differences between their genomes. He discovered the large-scale structure of the tree of life, with all living creatures descended from three primordial branches which he called bacteria, archea, and eukaryotes. He recently published a provocative and illuminating article with the title "A New Biology for a New Century" in the June 2004 issue of *Microbiology and Molecular Biology Reviews.* I am indebted to my biologist friend Gerald Joyce for bringing it to my attention. Woese's main theme is the obsolescence of reductionist biology as it has been practiced for the last hundred years, and the need for a new synthetic biology based on emergent patterns of organization rather than on genes and molecules. Aside from his main theme, he raises another important question. When did Darwinian evolution begin? By Darwinian evolution he means evolution as Darwin understood it, based on the competition for survival of non-interbreeding species. He presents evidence that Darwinian evolution did not go back to the beginning of life. When we compare genomes of ancient lineages of living creatures, we find evidence of massive transfers of genetic information from one lineage to another. In early times, horizontal gene transfer, the sharing of genes between unrelated species, was prevalent. It becomes more prevalent the further back you go in time.

Whatever Carl Woese writes, even in a speculative vein, needs to be taken seriously. In his "New Biology" article, he is postulating a golden age of pre-Darwinian life, when horizontal gene transfer was universal and separate species did not exist. Life was then a community of cells of various kinds, sharing their genetic information so that clever chemical tricks and catalytic processes invented by one creature could be inherited by all of them. Evolution was a communal affair, the whole community advancing in metabolic and reproductive efficiency as the genes of the most efficient cells were shared. Evolution could be rapid, as new chemical devices could be evolved simultaneously by cells of different kinds working in parallel and then reassembled in a single cell by horizontal gene transfer. But then, one evil day, a cell resembling a primitive bacterium happened to find itself one jump ahead of its neighbors in efficiency. That cell, anticipating Bill Gates by three billion years, separated itself from the community and refused to share. Its offspring became the first species of bacteria, reserving their intellectual property for their own private use. With their superior efficiency, the bacteria continued to prosper and to evolve separately, while the rest of the community continued its communal life. Some millions of years later, another cell separated itself from the community and became the ancestor of the archea. Some time after that, a third cell separated itself and became the ancestor of the eukaryotes. And so it went on, until nothing was left of the community and all life was divided into species. The Darwinian interlude had begun.

The Darwinian interlude has lasted for two or three billion years. It probably slowed down the pace of evolution considerably. The basic biochemical machinery of life had evolved rapidly during the few hundreds of millions of years of the pre-Darwinian era, and changed very little in the next two billion years of microbial evolution. Darwinian evolution is slow because individual species once established evolve very little. Darwinian evolution requires established species to become extinct so that new species can replace them.

Now, after three billion years, the Darwinian interlude is over.

It was an interlude between two periods of horizontal gene transfer. The epoch of Darwinian evolution based on competition between species ended about ten thousand years ago, when a single species, *Homo sapiens,* began to dominate and reorganize the biosphere. Since that time, cultural evolution has replaced biological evolution as the main driving force of change. Cultural evolution is not Darwinian. Cultures spread by horizontal transfer of ideas more than by genetic inheritance. Cultural evolution is running a thousand times faster than Darwinian evolution, taking us into a new era of cultural interdependence which we call globalization. And now, as *Homo sapiens* domesticates the new biotechnology, we are reviving the ancient pre-Darwinian practice of horizontal gene transfer, moving genes easily from microbes to plants and animals, blurring the boundaries between species. We are moving rapidly into the post-Darwinian era, when species will no longer exist, and the rules of Open Source sharing will be extended from the exchange of software to the exchange of genes. Then the evolution of life will once again be communal, as it was in the good old days before separate species and intellectual property were invented.

I would like to borrow Carl Woese's vision of the future of biology and extend it to the whole of science. Here is Carl Woese's metaphor for the future of science:

> Imagine a child playing in a woodland stream, poking a stick into an eddy in the flowing current, thereby disrupting it. But the eddy quickly reforms. The child disperses it again. Again it reforms, and the fascinating game goes on. There you have it! Organisms are resilient patterns in a turbulent flow—patterns in an energy flow. . . . It is becoming increasingly clear that to understand living systems in any deep sense, we must come to see them not materialistically, as machines, but as stable, complex, dynamic organization.

This picture of living creatures, as patterns of organization rather than collections of molecules, applies not only to butterflies and rain forests but also to thunderstorms and hurricanes. The non-living universe is as diverse and as dynamic as the living universe, and is also dominated by patterns of organization that are not yet understood. The reductionist physics and the reductionist molecular biology of the twentieth century will continue to be important in the twenty-first century, but they will not be dominant. The big problems, the evolution of the universe as a whole, the origin of life, the nature of human consciousness, and the evolution of the earth's climate, cannot be understood by reducing them to elementary particles and molecules. New ways of thinking and new ways of organizing large databases will be needed.

Green Technology

The domestication of biotechnology may also be helpful in solving practical economic and environmental problems. Once a new generation of children has grown up, as familiar with biotech games as our grandchildren are now with computer games, biotechnology will no longer seem weird and alien. In the era of Open Source biology, the magic of genes will be available to anyone with the skill and imagination to use it. The way will be open for biotechnology to move into the mainstream of economic development, to help us solve some of our urgent social problems and ameliorate the human condition all over the earth. Open Source biology could be a powerful tool, giving us access to cheap and abundant solar energy.

A plant is a creature that uses the energy of sunlight to convert water and carbon dioxide and other simple chemicals into roots and leaves and flowers. To live, it needs to collect sunlight. But it uses sunlight with low efficiency. The most efficient crop plants, such as sugarcane or maize, convert about 1 percent of the sunlight that falls onto them into chemical energy. Artificial solar collectors made of silicon can do much better. Silicon solar cells can convert sunlight into electrical energy with 15 percent efficiency, and electrical energy can be converted into chemical energy without much loss. We can imagine that in the future, when we have mastered the art of genetically engineering plants, we may breed new crop plants that have leaves made of silicon, converting sunlight into chemical energy with ten times the efficiency of natural plants. These artificial crop plants would reduce the area of land needed for biomass production by a factor of ten. They would allow solar energy to be used on a massive scale without taking up too much land. They would look like natural plants except that their leaves would be black instead of green. The question I am asking is, how long will it take us to grow plants with black leaves?

If the natural evolution of plants had been driven by the need for high efficiency of utilization of sunlight, then the leaves of all plants would have been black. Obviously the evolution was driven by other needs, and in particular by the need for protection against overheating. For a plant growing in a hot climate, it is advantageous to reflect as much as possible of the sunlight that is not used for growth. There is plenty of sunlight, and it is not important to use it with maximum efficiency. The plants have evolved with chlorophyll in their leaves to absorb the useful red and blue components of sunlight and to reflect the green. That is why it is reasonable for plants in tropical climates to be green. But this logic does not explain why plants in cold climates where sunlight is scarce are also green. We could imagine that in places like Alaska or Iceland, overheating would not be a problem, and plants with black leaves using sunlight more efficiently would have an evolutionary advantage. For some reason which we do not understand, natural plants with black leaves never appeared. Why not? Perhaps we shall not understand why nature did not travel this route until we have traveled it ourselves.

After we have explored this route to the end, when we have created new forests of black-leaved plants that can use sunlight ten times more efficiently than natural plants, we shall be confronted by a new set of environmental problems. Who shall be allowed to grow the black-leaved plants? Will black-leaved plants remain an artificially maintained cultivar, or will they invade and permanently change the natural ecology? What shall we do with the silicon trash that these plants leave behind them? Shall we be able to design a whole ecology of silicon-eating microbes and fungi and earthworms to keep the black-leaved plants in balance with the rest of nature and to recycle their silicon? The twenty-first century will bring us powerful new tools of genetic engineering with which to manipulate our farms and forests. With the new tools will come new questions and new responsibilities.

Rural poverty is one of the great evils of the modern world. The lack of jobs and economic opportunities in villages drives millions of people to migrate from villages into overcrowded cities. The continuing migration causes immense social and environmental problems in the major cities of poor countries. The effects of poverty are most visible in the cities, but the causes of poverty lie mostly in the villages. What the world needs is a technology that directly attacks the problem of rural poverty by creating wealth and jobs in the villages. A technology that creates industries and careers in villages would give the villagers a practical alternative to migration. It would give them a chance to survive and prosper without uprooting themselves.

The shifting balance of wealth and population between villages and cities is one of the main themes of human history over the last ten thousand years. The shift from villages to cities is strongly coupled with a shift from one kind of technology to another. I find it convenient to call the two kinds of technology green and grey. The adjective green has been appropriated and abused by various political movements, especially in Europe, so I need to explain clearly what I have in mind when I speak of green and grey. Green technology is based on biology, grey technology on physics and chemistry. Roughly speaking, green technology is the technology that gave birth to village communities ten thousand years ago, starting from the domestication of plants and animals, the invention of agriculture, the breeding of goats and sheep and horses and cows and pigs, the manufacture of textiles and cheese and wine. Grey technology is the technology that gave birth to cities and empires five thousand years later, starting from the forging of bronze and iron, the invention of wheeled vehicles and paved roads, the building of ships and war chariots, the manufacture of swords and guns and bombs. For the first five of the ten thousand years, wealth and power belonged to villages with green technology, and for the second five thousand years wealth and power belonged to cities with grey technology. Beginning about five hundred years ago, grey technology became increasingly dominant, as we learned to build machines using power from wind and water and steam and electricity. In the last hundred years, wealth and power were even more heavily concentrated in cities as grey technology raced ahead. As cities became richer, rural poverty deepened.

This sketch of the last ten thousand years of human history puts the problem of rural poverty into a new perspective. If rural poverty is a consequence of the unbalanced growth of grey technology, it is possible that a shift in the balance back from grey to green might cause rural poverty to disappear. That is my dream. During the last fifty years we have seen explosive progress in the scientific understanding of the basic processes of life, and in the last twenty years this new understanding has given rise to explosive growth of green technology. The new green technology allows us to breed new varieties of animals and plants as our ancestors did ten thousand years ago, but now a hundred times faster, taking a decade instead of a millennium to create a new crop plant. Guided by a precise understanding of genes and genomes instead of by trial and error, we can within a few years modify plants so as to give them improved yield, improved nutritive value, or improved resistance to pests and diseases.

Within a few more decades, as the continued exploring of genomes gives us better knowledge of the architecture of living creatures, we shall be able to design new species of microbes and plants according to our needs. The way will then be open for green technology to do more cheaply and more cleanly many of the things that grey technology can do, and also to do many things that grey technology has failed to do. Green technology could replace most of our existing chemical industries and a large part of our mining and manufacturing industries. Green technology could achieve more extensive recycling of waste products and worn-out machines, with great benefit to the environment. An economic system based on green technology could come much closer to the goal of sustainability, using sunlight instead of fossil fuels as the primary source of energy. New species of termite could be engineered to chew up derelict automobiles instead of houses, and new species of tree could be engineered to convert carbon dioxide and sunlight into liquid fuels instead of cellulose.

Before genetically modified termites and trees can be allowed to help solve our economic and environmental problems, great arguments will rage over the possible damage they may do. Many of the people who call themselves green are passionately opposed to green technology. But in the end, if the technology is developed carefully and deployed with sensitivity to human feelings, it is likely to be accepted by most of the people who will be affected by it, just as the equally unnatural and unfamiliar green technologies of milking cows and plowing soils and fermenting grapes were accepted by our ancestors long ago. I am not saying that the political acceptance of green technology will be quick or easy. I say only that green technology has enormous promise for preserving the balance of nature on this planet as well as for relieving human misery. Future generations of people raised from childhood with biotech toys and games will probably accept it more easily than we do. Nobody can predict how long it may take to try out the new technology in a thousand different ways and measure its costs and benefits.

What has this dream of a resurgent green technology to do with the problem of rural poverty? In the past, green technology has always been rural, based in farms and villages rather than in cities. In the future it will pervade cities as well as countryside, factories as well as forests. It will not be entirely rural. But it will still have a large rural component. After all, the cloning of Dolly occurred in a rural animal-breeding station in Scotland, not in an urban laboratory in Silicon Valley. Green technology will use land and sunlight as its primary sources of raw materials and energy. Land and sunlight cannot be concentrated in cities but are spread more or less evenly over the planet. When industries and technologies are based on land and sunlight, they will bring employment and wealth to rural populations. It is fortunate that sunlight is most abundant in tropical countries, where a large fraction of the world's people live and where rural poverty is most acute. Since sunlight is distributed more equitably than coal and oil, green technology can be a great equalizer, helping to narrow the gap between rich and poor countries.

My book *The Sun, the Genome, and the Internet* (1999) describes a vision of green technology enriching villages all over the world and halting the migration from villages to megacities. The three components of the vision are all essential: the sun to provide energy where it is needed, the genome to provide plants that can convert sunlight into chemical fuels cheaply and efficiently, the Internet to end the intellectual and economic isolation of rural populations. With all three components in place, every village in Africa could enjoy its fair share of the blessings of civilization. People who prefer to live in cities would still be free to move from villages to cities, but they would not be compelled to move by economic necessity.

Moore's Law in Biology

There are two kinds of physicists who dabble in biology. Some of us are arrogant and others are humble. Arrogant physicists say that biology needs better concepts, and since physicists are good at concepts, our job is to tell the biologists how to think. Humble physicists say that biology needs better hardware, and since physicists are good at hardware, our job is to invent new tools for biologists to use. Physicists who invent new tools have had the greater impact. Physicists invented optical phase-contrast and electron microscopes, X-ray diffraction and magnetic resonance imaging machines, besides the ubiquitous computers that biologists as well as physicists use to record and interpret experimental data.

Of all the tools created by physicists with grey technology, the most spectacularly successful is the integrated circuit, a little chip of silicon with a fabulous array of electronic circuits etched into its surface. For forty years the performance of integrated circuits has improved with time according to Moore's Law. Moore's Law says that their speed doubles every eighteen months, or increases by a factor of a hundred every decade, without much increase in cost. Computers today can do about a hundred million times as many operations per second as they could when Moore announced his law forty years ago, while the cost of a computer has remained roughly constant.

Moore's Law is sometimes true for the products of green technology too. Consider for example one of the central tools of green technology, the sequencing of DNA. Fred Sanger sequenced the first complete virus genome with five thousand base pairs in 1977, and the human genome with three billion base pairs was sequenced twenty-five years later. The output of base pairs followed Moore's Law, but the cost of sequencing did not. The human genome cost a great deal more than the virus genome. The sequencing machines that now exist are marvels of ingenuity, but they are cumbersome and expensive. They handle DNA molecules in bulk, using the methods of wet chemistry. The chemical reagents cost as much as the machines. What biology now needs is a single-molecule sequencer that can handle one molecule at a time and sequence it by physical rather than chemical methods. To invent such a machine is a job for physicists, using grey technology as a tool to support green technology. Whoever invents it and gets it to work reliably will make a major contribution to biology.

A single-molecule sequencer could be much cheaper as well as faster than existing sequencers. It might be as small and convenient as a laptop computer, zipping along a molecule of DNA as quickly as the polymerase enzyme that converts a single strand of DNA into a double strand, reading out base pairs into computer memory at a rate of a thousand per second. At that rate, a single machine could read out a complete human genome in a month. With plenty of hard work and a little luck, we shall evolve single-molecule sequencers that extend Moore's Law into the future, increasing the speed of sequencing and decreasing the unit cost by a factor of a hundred every decade.

What will this mean for green technology? Up to now we have sequenced genomes of about a hundred species, most of them microbes, with a total of about ten billion base pairs. The biosphere of our planet contains about ten million species, and their genomes contain altogether about ten quadrillion base pairs. In the language of computer science, the genomes of all the species on Earth add up to a few million gigabytes of data. This would be a database comparable in size with other databases that already exist. We already know how to store and search databases of this size. But before these genomes can be sequenced, the biosphere must be explored and the species identified. If Moore's Law remains valid for sequencing DNA, we can sequence the entire biosphere in about thirty years, at a cost not much greater than the cost of the human genome. The biosphere genome project will bring us to the beginning of a deep understanding of the biosphere, just as the human genome project has brought us to the beginning of a deep understanding of human biology. The sequencing of ten million species will be a good beginning, both for understanding and for preserving the biosphere. If we understand what is there, we shall have a better chance of preserving it.

ORIGINS OF LIFE

Carl Woese's vision of life as a dynamic process rather than a static collection of molecules throws some new light on an old mystery, the origins of life. An important clue to the understanding of the mystery is the dual nature of life as it now exists, with the two components, proteins and nucleic acids, looking like a symbiosis of two alien species. Proteins dominate the metabolism of modern cells and nucleic acids dominate their replication. Proteins are the busy little workers who ceaselessly scurry around the cell, taking the appropriate actions to keep the cell in good repair. Nucleic acids are the permanent handbooks of instructions, telling the cell how to copy itself precisely when it produces offspring. We do not know which component came first. The prevailing dogma among biologists says that nucleic acids came first. According to the dogma, there was an "RNA world" with creatures composed of RNA, another nucleic acid similar to DNA, before there were creatures composed of protein. This dogma received powerful support when the structure of the ribosome was determined. The ribosome is the chemical factory in which proteins are manufactured in every modern cell, making use of the instructions provided by nucleic acids. As a result of the recent discovery of the structure of the ribosome, we now know that its central core is made of RNA. Since the ribosome is probably the oldest surviving relic of ancestral forms of life, the fact that it is made of RNA strengthens the case for believing that RNA came first. The chief evidence against this belief is the fact that RNA is chemically unstable and hard to synthesize in a prebiotic environment.

Another key piece of evidence in the search for life's origins is the molecule adenosine triphosphate (ATP). This molecule is an essential link between the two components of living cells. ATP plays a universal role as a carrier of chemical energy in the metabolism of modern cells, and it is easily converted into a building block of nucleic acids. The question then is, which of the two roles of ATP came first? My answer to this question is, again, listen to the chemists. The chemists tell us that ATP, like the nucleic acids to which it is related, is chemically unstable and hard to synthesize without a preexisting metabolic apparatus. So I consider it most likely that the metabolic apparatus of primitive cells came first. I call this idea the garbage-bag model of the origin of life, with primitive cells containing a random mixture of organic chemicals, the cells growing and splitting into two from time to time but not replicating themselves precisely. According to this story, ATP appeared in preexisting garbage-bag cells as a helpful adjunct to their chemical machinery. And then, at a later stage of evolution, the building blocks derived from ATP began to assemble into nucleic acids and to develop their capacity for exact replication. Self-replicating RNA became a new parasitic form of life evolving inside preexisting cells. And then, after a while, the parasite became a symbiont, and life as we know it with its dual structure came into being.

This story of nucleic acids evolving as parasites within an older generation of cells is pure guesswork. I am not claiming that it is true. I claim only that it is a useful working hypothesis to explain the dual nature of life. The truth will certainly turn out to be more complicated. One consequence of the hypothesis is that the early parasitic phase of nucleic acid evolution was in fact the "RNA world." The RNA world existed inside preexisting cells. In the RNA world, RNA used its capacity for replication to evolve far more elaborate and fine-tuned chemical machinery than had previously existed in the primitive cells in which the RNA was housed. During this parasitic phase, RNA was on its own, decoupled from the metabolic apparatus of the cell. During this phase, RNA evolved its own metabolic machinery of RNA enzymes and other catalytic molecules, culminating in the evolution of the ribosome. Then the ribosome gave RNA the power to synthesize proteins, the proteins took over the organization of the cells, the nucleic acid apparatus and the protein apparatus were tied together, and the RNA world came to an end.

To obtain evidence for or against this hypothetical story of life's origins, a good strategy would be to explore the archaeology of the ribosome. The ribosomes of different species are all cousins, but they come in many different varieties. They are undeniably ancient, and their similarities and differences will provide detailed information about the chemical environment in the era when the RNA world was coming to an end. Perhaps, when the evidence buried in the structures and functions of the various lineages of ribosomes is deciphered, it will tell us whether they evolved inside preexisting cells or as independent free-living structures.

Woese's image of the child poking a stick into an eddy, his picture of life as a resilient pattern in a turbulent flow, is particularly helpful when we are thinking about life's origins. From the beginning, life was dynamic rather than static. When we look at the two components in every living cell today, we see a dynamic assemblage of proteins and smaller molecules supporting a comparatively static assemblage of nuclear acids. Since I find Woese's picture of life persuasive, I find it natural to think of the dynamic elements of life coming first, the static elements second. That means that metabolism came first, exact replication second. If this picture of life's origins is correct, then there was indeed an RNA world, but the world of primitive cells with largely random metabolism and without replication came first, the RNA world came second, and the modern world of life with metabolism tied to replication came third.

2
A Debate with Bill Joy

Invitation to the Magic Mountain

Biotechnology is likely to be the main driving force of change in human affairs for the next hundred years. I am an unashamed optimist, and I see the promise of good arising from biotechnology greatly outweighing the dangers of evil. But I am well aware that not everyone agrees with me about this. I try to keep the discussion balanced, with the pessimists also having their say. The subject of this chapter is a debate between Bill Joy and me. Bill Joy is a thoughtful entrepreneur who made a fortune in the computer industry and is now arguing that people like himself are too dangerous to be allowed. He makes a strong case for restraint of new technologies, and I do my best to refute him. Whether you agree with him or with me, I hope you will find our debate illuminating.

In January 2001 I was invited to the World Economic Forum in Davos, Switzerland. Most of the people there are captains of industry, presidents of foundations, or government officials. But in 2001 they decided to invite a few scientists and writers and artists to give the meeting a bit of intellectual tone, and I was one of the lucky ones. I was particularly lucky to be invited in 2001 and not in 2002, because the 2002 meeting was held in New York, and Davos is much more fun than New York. Davos is the Magic Mountain that Thomas Mann wrote about in his famous novel. In the novel, the characters are patients in a tuberculosis sanatorium who sit up there on the mountain and talk about the state of their souls. That happened almost a hundred years ago, before the First World War. The joke is that almost nothing has changed. The sanatorium is still there, and my wife's nephew works there happily as a physician. He can go skiing every day after he is done with the patients. The patients are still there. They still talk about the state of their souls. The only thing that has changed is the disease. The fashionable disease for patients who can afford to stay at the sanatorium is now not tuberculosis but asthma. Asthma has the advantage of being incurable, so the patients keep on coming back.

For the week of the meeting, Davos is full of celebrities. It is a name-dropper's paradise. At lunch we talked with Ian Wilmut, the man who cloned Dolly. At dinner we talked with Reinhold Messner, the man who climbed Everest solo without oxygen and is now a member of the European Parliament. We did not talk with Bill Gates or Yasser Arafat or President Mbeki of South Africa, but we saw them and heard them speak. The meeting is a great chance for important people to get to know each other, and that is why they come. Unimportant people like me are invited to provide some incidental entertainment. The entertainment that I was asked to provide was a public debate with Bill Joy, founder and chief scientist at Sun Microsystems, a large and successful computer company, on the question, Is our technology out of control? Bill was taking an extreme position on the yes side, and I was invited to take an extreme position on the no side, to make the debate interesting. Luckily I have reliable documentation of the debate to supplement my unreliable memory. Bill and I had conducted our debate in writing a few months before the Davos meeting. Our debate in Davos was a fake, a rehearsed performance, and perhaps for that reason it was a great success, and we had to repeat it in a larger hall.

In addition to our debate, there was another debate at Davos, on the subject of genetically modified (GM) food, that raised many of the same issues. I was a listener and not a speaker at the GM food debate, and I found it very illuminating. It was a debate between Europe and Africa. The Europeans oppose GM food with religious zeal. They say it is destroying the balance of nature, with unacceptable risks to human health and natural ecology. They talked a great deal about a rule called the Precautionary Principle. The Precautionary Principle says that if some course of action carries even a remote chance of irreparable damage to the ecology, then you shouldn't do it, no matter how great the possible advantages of the action may be. You are not allowed to balance costs against benefits when deciding what to do. The Precautionary Principle gives the Europeans a firm philosophical basis for saying no to GM food.

In response, the Africans pointed out that the Precautionary Principle can just as well be used as a philosophical basis for saying yes. The growing population and general impoverishment of Africa are already causing irreparable damage to the ecology, and saying no to GM food will only make the irreparable damage worse. The European pretense of allowing no risk of irreparable damage makes no sense in the real world. In the real world there are risks of irreparable damage no matter what you do. There is no escape from balancing one risk against another. The Africans need GM crops in order to survive. In most of Africa, soils are poor, droughts are devastating, and many crops are lost to disease and pests. GM crops can make the difference between starving and surviving for subsistence farmers, between prosperity and ruin for cash farmers. Africans need to sell products to Europe. The European ban on GM food protects European farmers and hurts the Africans. As the Africans see it, the European ban on GM food is motivated more by economic advantage than by philosophical purity. I was happy to hear Bruce Alberts, the president of the U.S. National Academy of Sciences, supporting the Africans. Jerome Rifkin, an American activist who makes a career out of opposing new technologies, was supporting the Europeans. The arguments about GM food continue, with no end in sight.

Bill Joy Speaks

The day after the GM food debate, I had my debate with Bill Joy. Bill spoke first, following the line he had taken in an article he published in *Wired* magazine in April 2000 with the title "Why the Future Doesn't Need Us" (2000a). By "us" he means people like himself who develop and sell new technology. Since I do not remember his spoken words precisely, I quote passages from his writings which contain the gist of his argument. Here are some quotations from the *Wired* article.

First quotation:

> As society and the problems that face it become more and more complex and machines become more and more intelligent, people will let machines make more of their decisions for them, simply because machine-made decisions will bring better results than man-made ones. Eventually a stage may be reached at which the decisions necessary to keep the system running will be so complex that human beings will be incapable of making them intelligently. At that stage the machines will be in effective control. People won't be able to just turn the machines off, because they will be so dependent on them that turning them off would amount to suicide.

Now the surprise. This quotation was written not by Bill Joy but by Theodore Kaczynski, the Unabomber, who killed three people and wounded many others during a seventeen-year campaign of terror. Bill remarked that during those years he felt that he could easily have been the Unabomber's next victim. All the same, Bill said, the Unabomber had a valid point. Even if we are glad he is behind bars, we had better listen to what he is saying.

Second quotation. This is Bill himself writing:

> The twenty-first-century technologies, genetics, nanotechnology and robotics, are so powerful that they can spawn whole new classes of accidents and abuses. Most dangerously, for the first time, these accidents and abuses are widely within the reach of individuals or small groups. They will not require large facilities or rare raw materials. Knowledge alone will enable the use of them. Thus we have the possibility not just of weapons of mass destruction but of knowledge-enabled mass destruction, their destructiveness hugely amplified by the power of self-replication. I think it is no exaggeration to say we are on the cusp of the further perfection of extreme evil, an evil whose possibility spreads well beyond that which weapons of mass destruction bequeathed to the nation-states, on to a surprising and terrible empowerment of extreme individuals.

This was written a year and a half before the events of September 2001. I don't know whether Bill at that time had Osama bin Laden as well as the Unabomber in mind. He certainly had in mind the possibility of a Unabomber taking his revenge on society with genetically engineered microbes rather than with chemical explosives.

Third quotation. Here Bill is quoting from Eric Drexler, the chief prophet of nanotechnology. Drexler set up the Foresight Institute, which has now been in existence for twenty-five years, to promote the benign uses of nanotechnology and to warn against the dangerous uses. Here is Drexler:

> Artificial plants, with artificial leaves no more efficient than today's solar cells, could out-compete real plants, crowding the biosphere with inedible foliage. Tough omnivorous artificial bacteria could out-compete real bacteria. They could spread like blowing pollen, replicate swiftly, and reduce the biosphere to dust in a matter of days. Dangerous replicators could easily be too tough, small, and rapidly spreading to stop, at least if we make no preparation. We have trouble enough controlling viruses and fruit flies. Among the cognoscenti of nanotechnology, this threat has become known as the gray goo problem. Though masses of uncontrolled replicators need not be gray or gooey, the term gray goo emphasizes that replicators able to obliterate life might be less inspiring than a single species of crabgrass. They might be superior in an evolutionary sense, but this need not make them valuable. The gray goo threat makes one thing perfectly clear: We cannot afford certain kinds of accidents with replicating assemblers.

Let me explain what Drexler means by replicating assemblers. The idea of nanotechnology is to build machines on a tiny scale that are as capable as living cells but made of different materials so that they are more rugged and more versatile. One kind of nanomachine is the assembler, which is a tiny factory that can manufacture other machines, including replicas of itself. Drexler understood from the beginning that a replicating assembler would be a tool of immense power for good or for evil. Fortunately or unfortunately, nanotechnology has moved much more slowly than Drexler expected, and nothing remotely resembling an assembler has yet emerged. The most useful products of nanotechnology so far are new composite materials, new kinds of computer chips, and MEMS. MEMS are Miniature Electro-mechanical Systems. These are special-purpose devices, mostly used to manipulate small quantities of fluids for chemical analysis or medical diagnosis. They have no capacity for replicating either themselves or anything else.

My last quotation comes from an article Bill published in the *Washington Post* (2000b) summing up the dangers he foresees and recommending a program of action to avoid disasters. Fourth quotation:

> The nearest-term danger is the release of a deadly pathogen: a bioengineered white plague that could be highly infectious, have a long incubation period and be targeted on specific groups. Nanotechnology poses the threat of a gray goo, engineered from materials foreign to the environment, which would outcompete the existing biosphere. In both cases, recall of these menaces would be impossible. A global disaster could occur in weeks. . . .
>
> We who are involved in advancing the new technologies must devote our best efforts to heading off disaster. I offer here a list of first steps suggested by our history with weapons of mass destruction: (1) Have scientists and technologists, and corporate leaders as well, take a vow, along the lines of the Hippocratic Oath, to avoid work on potential and actual weapons of mass destruction. (2) Create an international body to publicly examine the dangers and ethical issues of new technology. (3) Use stricter notions of liability, forcing companies to take responsibility for consequences through a private-sector mechanism, insurance. (4) Internationalize control of knowledge and technologies that have great potential but are judged too dangerous to be made commercially available. (5) Relinquish pursuit of that knowledge, and relinquish development of those technologies, so dangerous that we judge it better that they never be available. I too believe in the pursuit of knowledge and development of technologies. Yet we already have seen cases, such as biological weapons, where relinquishment is the obvious wise choice.

Dyson Responds

Bill then sat down, and I responded. I agreed that the dangers he describes are real, but I disagreed with some details of his argument, and I disagreed strongly with his remedies. My argument is divided into four parts. First, I say that the main dangers come from biotechnology and not from nanotechnology. Second, the dangers are not new and have been taken seriously by biologists and public health professionals for many years. Third, he ignores the long history of effective action by the international biological community to regulate and prohibit dangerous technologies. Fourth, I agree that we can relinquish the pursuit of dangerous new technologies, by the means that he suggests or otherwise, but to relinquish the pursuit of dangerous knowledge is neither possible nor desirable.

I took these four topics in turn. First, I spoke about nanotechnology. Thirty years ago, when Eric Drexler began to promote nanotechnology as the wave of the future, nanotechnology and biotechnology were emerging together and appeared to be about even in their stage of development. It seemed then that nanotechnology, constructing small machines by mechanical manipulation, and biotechnology, constructing small machines by genetic manipulation, were running side by side in the race to become practical and useful. Since that time, biotechnology has raced ahead with the new tools of gene splicing and polymerase chain reaction, while nanotechnology has moved ahead more slowly.

One of the goals of nanotechnology is to mass-produce tiny building blocks that might be assembled into tiny machines. A few years ago, Gerald Joyce and his friends at the Scripps Research Institute in California published a paper in *Nature* announcing the first mass production of such building blocks (Shih, Quispe, and Joyce 2004). They showed pictures of building blocks in the shape of regular octahedra, identical in size and structure, each of them perfectly rigid with a diameter of twenty-two nanometers. They look as if they were solid objects shaped and sliced by a mechanical tool. But in fact they are nothing of the kind. They are made of pure DNA and are synthesized by the polymerase chain reaction. The technology that produced them is pure biotechnology and has nothing to do with machine tools. Joyce and his friends designed a DNA molecule so that it would fold spontaneously into the octahedral shape as a result of bonding of complementary DNA bases. They mass-produced it by inserting the DNA into the genome of E. coli bacteria and letting the bacteria multiply. With a little help from E. coli, a group of biologists beat the nanotechnologists at their own game.

What would it take to construct the self-replicating assembler that Eric Drexler proclaimed as the holy grail of nanotechnology twenty-five years ago? The mathematician John von Neumann answered this question definitively in 1948 in his classic study of mechanical automata (Von Neumann 1948). He proved by logical analysis that any self-replicating automaton must have four components: Component A is an automatic factory. Component B is a copying machine. Component C is a control machine to control the actions of A and B. Component D is a blueprint containing a complete description of A and B and C. Five years later, Crick and Watson discovered the double helix, and it became clear that every living cell possesses these four components. A is the ribosomes, B is the polymerase enzymes, C is the enzymes controlling cell division, D is the genome composed of DNA or RNA. Now suppose that the nanotechnologists are trying to construct a self-replicating assembler. They will have to construct these four components and arrange them to function together as they do in a living cell. In the end they will only have reinvented something very similar to a bacterium. Biotechnology can do the same job more reliably and much more cheaply. This being so, the menace of grey goo loses credibility. If the self-replicating assembler is made of truly nonbiological materials, it will not be able to replicate itself by eating existing forms of life. If it is made of biological materials, then it is only another kind of biological disease germ. Either way, the nightmare of the assembler becoming a grey goo and eating up the biosphere cannot happen. It may be very nasty, but it is not going to wipe us out. I conclude that the dangers of nanotechnology, like its achievements, have been oversold. The dangers, if it succeeds in achieving its goals, will be essentially the same as the dangers of biotechnology.

Second, I spoke about the biologists who have worried about the dangers of biotechnology for at least forty years and have successfully taken political action to keep disasters from happening. I spoke about Matthew Meselson and Joshua Lederberg, leaders in the fight against biological weapons, and Maxine Singer and Paul Berg, leaders in the fight for international regulation of gene-splicing experiments. As a result of their efforts, the international community of biologists, and to some extent the international community of political leaders, are aware of the magnitude of the dangers and of the need for further action as the technology moves ahead. Nobody denies that biological weapons are an enormous threat, and that the threat comes as much from freelance terrorist groups as from national governments. But the biologists have been far more wholehearted and effective in fighting against biological weapons than physicists have been in fighting against nuclear weapons. Concerned citizens like Bill Joy who are new to the game have a lot to learn from the biologists.

Third, I spoke about the history of biological weapons and gene-splicing experiments, and the successes and failures of efforts to regulate them. Gene-splicing experiments began in many countries when the technique of sticking pieces of DNA together was discovered in 1975. Maxine Singer and Paul Berg issued a call for a moratorium on all such experiments until the dangers could be carefully assessed. There were obvious dangers to public health if genes for deadly toxins could be inserted into bacteria that are normally endemic in human populations. Biologists all over the world quickly agreed to the moratorium, and experiments were halted everywhere for ten months. During the ten months, two international conferences were held to work out the guidelines for permissible and forbidden experiments. The guidelines established rules of physical and biological containment for permitted experiments involving various degrees of risk. The most dangerous experiments were forbidden outright. These guidelines were adopted voluntarily by biologists and have been observed ever since, with some changes made from time to time in response to new discoveries. As a result, no serious health hazards have arisen from the experiments in twenty-five years. This is a shining example of responsible citizenship, showing that it is possible for scientists to protect the public from injury while preserving the freedom of science.

The history of biological weapons is a more complicated story. The United States, Britain, and the Soviet Union all had large programs to develop and stockpile biological weapons during and after the Second World War. But these were low-key efforts compared with the programs to develop nuclear weapons. Unlike the leading physicists who pushed the nuclear bomb programs ahead with great enthusiasm, the leading biologists never pushed hard for biological weapons. The great majority of biologists had nothing to do with weapons. The few leading biologists who were involved with the weapons program were mostly opposed to it. The strongest of the opponents in the United States was Matthew Meselson, who had the good luck to be a neighbor and friend of Henry Kissinger in 1968 when Nixon became president. Kissinger became national security adviser to President Nixon. Meselson seized the opportunity to convince Kissinger, and Kissinger convinced Nixon, that the American biological weapons program was far more dangerous to the United States than to any possible enemy. On the one hand, it was difficult to imagine any circumstances in which the United States would wish to use these weapons, and on the other hand, it was easy to imagine circumstances in which some of the weapons could fall into the hands of a group of terrorists. So Nixon in 1969 boldly declared that the United States was dismantling the entire program and destroying the stockpile of weapons. This was a unilateral move, not requiring any international agreement or ratification by the American

Senate. The development of weapons was duly stopped, and the weapons were destroyed. Britain quickly followed suit. In 1972, as a result of Nixon's initiative, an international convention was signed by the United States, the United Kingdom, and the Soviet Union imposing a permanent prohibition of biological weapons on all three countries. Many other countries subsequently signed the convention.

As we now know, the Soviet Union violated the biological weapons convention of 1972 on a massive scale, continuing to develop new weapons and to accumulate stockpiles until the collapse of the USSR in 1991. After the collapse, Russia declared its adherence to the convention and announced that the Soviet program had now finally been stopped. But many of the old Soviet research and production centers remain hidden behind walls of secrecy, and Russia has never provided the world with convincing evidence that the program is not continuing. It is quite possible that stockpiles of biological weapons continue to exist in Russia and in other countries. Nevertheless, the 1972 convention remains legally in force, and the great majority of countries have signed it. Although the convention is unverifiable and although it has been violated, we are far better off with it than without it. Without the convention, we would not have any legal ground for complaint or for preventive action whenever a biological weapons program anywhere in the world is discovered. With the convention, the danger of biological weapons is not eliminated but it is significantly reduced. Again, biologists in general and Matthew Meselson in particular deserve credit for making this happen in the real world of national politics and international rivalries.

The fourth and last part of my reply to Bill Joy concerns the question of remedies for the dangers that we all agree exist. Bill says, "Internationalize control of knowledge," and "Relinquish pursuit of that knowledge . . . so dangerous that we judge it better that [it] never be available." He is advocating censorship of scientific inquiry, either by international or by national authorities. I am opposed to this kind of censorship, and I will now explain why. From this point on, my argument in the Davos debate followed the same script that I had used when I testified on a previous occasion to the U.S. Congress.

In 1977, after the biologists had established their voluntary guidelines to reduce the dangers from gene-splicing experiments, many voices were heard saying the same things that Bill Joy is saying now, saying that voluntary guidelines were insufficient, saying that the dangers were so great that the pursuit of knowledge must be controlled by public authority. The subcommittee on Science, Research and Technology of the U.S. House of Representatives held hearings to discuss the dangers and the possible remedies. I was invited to testify, and this is what I said:

> It has sometimes been said that the risks of [gene-splicing] technology are historically unparalleled because the consequences of letting a new living creature loose in the world may be irreversible. I think we can find many historical parallels where governments were trying to guard against dangers that were equally irreversible.
>
> Three hundred and thirty-three years ago [it is now 362 years], the poet John Milton wrote a speech with the title "Areopagitica," addressed to the parliament of England. He was arguing for the liberty of unlicensed printing. I am suggesting that there is an analogy between the seventeenth-century fear of moral contagion by soul-corrupting books and the twentieth-century fear of physical contagion by pathogenic microbes. In both cases, the fear was neither groundless nor unreasonable. In 1644, when Milton was writing, England was engaged in a long and bloody civil war, and the Thirty Years' War which devastated Germany had still four years to run. These seventeenth-century wars were religious wars in which differences of doctrine played a great part. In that century, books not only corrupted souls but also mangled bodies. The risks of letting books

go free into the world were rightly regarded by the English parliament as potentially lethal as well as irreversible. Milton argued that the risks must nevertheless be accepted. I ask you to consider whether his message may still have value for our own times, if the word "book" is replaced by the word "experiment." Here is Milton:

> I deny not but that it is of greatest concernment in the Church and Commonwealth, to have a vigilant eye how books demean themselves as well as men, and thereafter to confine, imprison, and do sharpest justice on them as malefactors. I know they are as lively, and as vigorously productive, as those fabulous dragon's teeth, and being sown up and down, may chance to spring up armed men.

The important word in Milton's statement is "thereafter." Books should not be convicted and imprisoned until after they have done some damage. What Milton objected to was prior censorship, that books would be prohibited even from seeing the light of day. Next, Milton comes to the heart of the matter, the difficulty of regulating "things uncertainly and yet equally working to good and to evil."

> Suppose we could expel sin by this means; look how much we thus expel of sin, so much we expel of virtue: for the matter of them both is the same; remove that, and ye remove them both alike. This justifies the high providence of God, who, though he commands us temperance, justice, continence, yet pours out before us, even to a profuseness, all desirable things, and gives us minds that can wander beyond all limit and satiety. Why should we then affect a rigor contrary to the manner of God and of nature, by abridging or scanting those means, which books freely permitted are, both to the trial of virtue, and the exercise of truth? It would be better done, to learn that the law must needs be frivolous, which goes to restrain things, uncertainly and yet equally working to good and to evil.

Next I quote a passage about Galileo. It shows that the connection between the silencing of Galileo and the general decline of intellectual life in seventeenth-century Italy was obvious to a contemporary eyewitness.

> And lest some should persuade ye, Lords and Commons, that these arguments of learned men's discouragement at this your order are mere flourishes, and not real, I could recount what I have seen and heard in other countries, where this kind of inquisition tyrannizes; when I have sat among their learned men, for that honor I had, and been counted happy to be born in such a place of philosophic freedom, as they supposed England was, while themselves did nothing but bemoan the servile condition into which learning amongst them was brought; that this was it which had damped the glory of Italian wits; that nothing had been there written now these many years but flattery and fustian. There it was that I found and visited the famous Galileo, grown old, a prisoner to the Inquisition, for thinking in astronomy otherwise than the Franciscan and Dominican licencers thought.

My last quotation expresses Milton's patriotic pride in the intellectual vitality of seventeenth-century England, a pride that twenty-first-century Americans have good reason to share. "Lords and Commoners of England, consider what nation it is whereof ye are, and whereof ye are the governors; a nation not slow and dull, but of a quick, ingenious and piercing spirit, acute to invent, subtle and sinewy to discourse, not beneath the reach of any point the highest that human capacity can soar to. Nor is it for nothing that the grave and frugal Transylvanian sends out yearly from the mountainous borders of Russia, and beyond the Hercynian wilderness, not their youth, but their staid men, to learn our language and our theologic arts." Perhaps, after all, as we struggle to deal with the enduring problems of reconciling individual freedom with public safety, the wisdom of a great poet may be a surer guide than the calculations of risk-benefit analysis.

That was the end of my pitch. After that, the debate continued with statements and questions from the audience which I do not remember. No vote was taken to determine who won. The purpose of the debate was not to win but to educate. Bill Joy and I remain friends.

3
Heretical Thoughts about Science and Society

The Need for Heretics

In the modern world, science and society often interact in a perverse way. We live in a technological society, and technology causes political problems. The politicians and the public expect science to provide answers to the problems. Scientific experts are paid and encouraged to provide answers. The public does not have much use for a scientist who says, "Sorry, but we don't know." The public prefers to listen to scientists who give confident answers to questions and make confident predictions of what will happen as a result of human activities. So it happens that the experts who talk publicly about politically contentious questions tend to speak more clearly than they think. They make confident predictions about the future, and end up believing their own predictions. Their predictions become dogmas which they do not question. The public is led to believe that the fashionable scientific dogmas are true, and it may sometimes happen that they are wrong. That is why heretics who question the dogmas are needed.

As a scientist I do not have much faith in predictions. Science is organized unpredictability. The best scientists like to arrange things in an experiment to be as unpredictable as possible, and then they do the experiment to see what will happen. You might say that if something is predictable then it is not science. When I make predictions, I am not speaking as a scientist. I am speaking as a storyteller, and my predictions are science fiction rather than science. The predictions of science-fiction writers are notoriously inaccurate. Their purpose is to imagine what might happen rather than to describe what will happen. I will be telling stories that challenge the prevailing dogmas of today. The prevailing dogmas may be right, but they still need to be challenged. I am proud to be a heretic. The world always needs heretics to challenge the prevailing orthodoxies. Since I am a heretic, I am accustomed to being in the minority. If I could persuade everyone to agree with me, I would not be a heretic.

We are lucky that we can be heretics today without any danger of being burned at the stake. But unfortunately I am an old heretic. Old heretics do not cut much ice. When you hear an old heretic talking, you can always say, "Too bad he has lost his marbles," and pass on. What the world needs is young heretics. I am hoping that one or two of the people who read this book may fill that role.

Two years ago, I was at Cornell University celebrating the life of Tommy Gold, a famous astronomer who died at a ripe old age. He was famous as a heretic, promoting unpopular ideas that usually turned out to be right. Long ago I was a guinea pig in Tommy's experiments on human hearing. He had a heretical idea that the human ear discriminates pitch by means of a set of tuned resonators with active electromechanical feedback. He published a paper explaining how the ear must work (Gold 1948). He described how the vibrations of the inner ear must be converted into electrical signals which feed back into the mechanical motion, reinforcing the vibrations and increasing the sharpness of the resonance. The experts in auditory physiology ignored his work because he did not have a degree in physiology. Many years later, the experts discovered the two kinds of hair cells in the inner ear that actually do the feedback as Tommy had predicted, one kind of hair cell acting as electrical sensors and the other kind acting as mechanical drivers. It took the experts forty years to admit that he was right. Of course, I knew that he was right because I had helped him do the experiments.

Later in his life, Tommy Gold promoted another heretical idea, that the oil and natural gas in the ground come up from deep in the mantle of the earth and have nothing to do with biology. Again the experts are sure that he is wrong, and he did not live long enough to change their minds. Just a few weeks before he died, some chemists at the Carnegie Institution in Washington did a beautiful experiment in a diamond anvil cell (Scott et al. 2004). They mixed together tiny quantities of three things that we know exist in the mantle of the earth, and observed them at the pressure and temperature appropriate to the mantle about two hundred kilometers down. The three things were calcium carbonate, which is sedimentary rock; iron oxide, which is a component of igneous rock; and water. These three things are certainly present when a slab of subducted ocean floor descends from a deep ocean trench into the mantle. The experiment showed that they react quickly to produce lots of methane, which is natural gas. Knowing the result of the experiment, we can be sure that big quantities of natural gas exist in the mantle two hundred kilometers down. We do not know how much of this natural gas pushes its way up through cracks and channels in the overlying rock to form the shallow reservoirs of natural gas that we are now burning. If the gas moves up rapidly enough, it will arrive intact in the cooler regions where the reservoirs are found. If it moves too slowly through the hot region, the methane may be reconverted to carbonate rock and water. The Carnegie Institution experiment shows that there is at least a possibility that Tommy Gold was right and the natural gas reservoirs are fed from deep below. The chemists sent an e-mail to Tommy Gold to tell him their result and got back a message that he had died three days earlier. Now that he is dead, we need more heretics to take his place.

Climate and Land Management

The main subject of this chapter is the problem of climate change. This is a contentious subject, involving politics and economics as well as science. The science is inextricably mixed up with politics. Everyone agrees that the climate is changing, but there are violently diverging opinions about the causes of change, about the consequences of change, and about possible remedies. I am promoting a heretical opinion, the first of three heresies that I will discuss in this chapter.

My first heresy says that all the fuss about global warming is grossly exaggerated. Here I am opposing the holy brotherhood of climate model experts and the crowd of deluded citizens who believe the numbers predicted by the computer models. Of course, they say, I have no degree in meteorology and I am therefore not qualified to speak. But I have studied the climate models and I know what they can do. The models solve the equations of fluid dynamics, and they do a very good job of describing the fluid motions of the atmosphere and the oceans. They do a very poor job of describing the clouds, the dust, the chemistry, and the biology of fields and farms and forests. They do not begin to describe the real world that we live in. The real world is muddy and messy and full of things that we do not yet understand. It is much easier for a scientist to sit in an air-conditioned building and run computer models than to put on winter clothes and measure what is really happening outside in the swamps and the clouds. That is why the climate model experts end up believing their own models.

There is no doubt that parts of the world are getting warmer, but the warming is not global. I am not saying that the warming does not cause problems. Obviously it does. Obviously we should be trying to understand it better. I am saying that the problems are grossly exaggerated. They take away money and attention from other problems that are more urgent and more important, such as poverty and infectious disease and public education and public health, and the preservation of living creatures on land and in the oceans, not to mention easy problems such as the timely construction of adequate dikes around the city of New Orleans.

I will discuss the global warming problem in detail because it is interesting, even though its importance is exaggerated. One of the main causes of warming is the increase of carbon dioxide in the atmosphere resulting from our burning of fossil fuels such as oil and coal and natural gas. To understand the movement of carbon through the atmosphere and biosphere, we need to measure a lot of numbers. I do not want to confuse you with a lot of numbers, so I will ask you to remember just one number. The number that I ask you to remember is one-hundredth of an inch per year. Now I will explain what this number means. Consider the half of the land area of the earth that is not desert or ice cap or city or road or parking lot. This is the half of the land that is covered with soil and supports vegetation of one kind or another. Every year it absorbs and converts into biomass a certain fraction of the carbon dioxide that we emit into the atmosphere. Biomass means living creatures, plants and microbes and animals, and the organic materials that are left behind when the creatures die and decay. We don't know how big a fraction of our emissions is absorbed by the land, since we have not measured the increase or decrease of the biomass. The number that I ask you to remember is the increase in thickness, averaged over one-half of the land area of the planet, of the biomass that would result if all the carbon that we are emitting by burning fossil fuels were absorbed. The average increase in thickness is one-hundredth of an inch per year.

The point of this calculation is the very favorable rate of exchange between carbon in the atmosphere and carbon in the soil. To stop the carbon in the atmosphere from increasing, we only need to grow the biomass in the soil by a hundredth of an inch per year. Good topsoil contains about 10 percent biomass (Schlesinger 1977), so a hundredth of an inch of biomass growth means about a tenth of an inch of topsoil. Changes in farming practices such as no-till farming, avoiding the use of the plow, cause biomass to grow at least as fast as this. If we plant crops without plowing the soil, more of the biomass goes into roots that stay in the soil, and less returns to the atmosphere. If we use genetic engineering to put more biomass into roots, we can probably achieve much more rapid growth of topsoil. I conclude from this calculation that the problem of carbon dioxide in the atmosphere is a problem of land management, not a problem of meteorology. No computer model of atmosphere and ocean can hope to predict the way we shall manage our land.

Here is another heretical thought. Instead of calculating worldwide averages of biomass growth, we may prefer to look at the problem locally. Consider a possible future, with China continuing to develop an industrial economy based largely on the burning of coal, and the United States deciding to absorb the resulting carbon dioxide by increasing the biomass in our topsoil. The quantity of biomass that can be accumulated in living plants and trees is limited, but there is no limit to the quantity that can be stored in topsoil. To grow topsoil on a massive scale may or may not be practical, depending on the economics of farming and forestry. It is at least a possibility to be seriously considered that China could become rich by burning coal, while the United States could become environmentally virtuous by accumulating topsoil, with transport of carbon from mine in China to soil in America provided free of charge by the atmosphere, and the inventory of carbon in the atmosphere remaining constant. We should take such possibilities into account when we listen to predictions about climate change and fossil fuels. If biotechnology takes over the planet in the next fifty years, as computer technology has taken it over in the last fifty years, the rules of the climate game will be radically changed.

When I listen to the public debates about climate change, I am impressed by the enormous gaps in our knowledge, the sparseness of our observations, and the superficiality of our theories. Many of the basic processes of planetary ecology are poorly understood. They must be better understood before we can reach an accurate diagnosis of the present condition of our planet. When we are trying to take care of a planet, just as when we are taking care of a human patient, diseases must be diagnosed before they can be cured. We need to observe and measure what is going on in the biosphere, rather than rely on computer models.

Everyone agrees that the increasing abundance of carbon dioxide in the atmosphere has two important consequences, first a change in the physics of radiation transport in the atmosphere, and second a change in the biology of plants on the ground and in the ocean. Opinions differ on the relative importance of the physical and biological effects, and on whether the effects, either separately or together, are beneficial or harmful. The physical effects are seen in changes of rainfall, cloudiness, wind strength, and temperature, which are customarily lumped together in the misleading phrase *global warming.* In humid air, the effect of carbon dioxide on radiation transport is unimportant because the transport of thermal radiation is already blocked by the much larger greenhouse effect of water vapor. The effect of carbon dioxide is important where the air is dry, and air is usually dry only where it is cold. Hot desert air may feel dry but often contains a lot of water vapor. The warming effect of carbon dioxide is strongest where air is cold and dry, mainly in the Arctic rather than in the tropics, mainly in mountainous regions rather than in lowlands, mainly in winter rather than in summer, and mainly at night rather than in daytime. The warming is real, but it is mostly making cold places warmer rather than making hot places hotter. To represent this local warming by a global average is misleading.

The fundamental reason why carbon dioxide in the atmosphere is critically important to biology is that there is so little of it. A field of corn growing in full sunlight in the middle of the day uses up all the carbon dioxide within a meter of the ground in about five minutes. If the air were not constantly stirred by convection currents and winds, the corn would stop growing. About a tenth of all the carbon dioxide in the atmosphere is converted into biomass every summer and given back to the atmosphere every fall. That is why the effects of fossil-fuel burning cannot be separated from the effects of plant growth and decay. There are five reservoirs of carbon that are biologically accessible on a short timescale, not counting the carbonate rocks and the deep ocean, which are only accessible on a timescale of thousands of years. The five accessible reservoirs are the atmosphere, the land plants, the topsoil in which land plants grow, the surface layer of the ocean in which ocean plants grow, and our proved reserves of fossil fuels. The atmosphere is the smallest reservoir and the fossil fuels are the largest, but all five reservoirs are of comparable size. They all interact strongly with one another. To understand any of them, it is necessary to understand all of them.

As an example of the way different reservoirs of carbon dioxide may interact with each other, consider the atmosphere and the topsoil. Greenhouse experiments show that many plants growing in an atmosphere enriched with carbon dioxide react by increasing their root-to-shoot ratio. This means that the plants put more of their growth into roots and less into stems and leaves. A change in this direction is to be expected, because the plants have to maintain a balance between the leaves collecting carbon from the air and the roots collecting mineral nutrients from the soil. The enriched atmosphere tilts the balance so that the plants need less leaf area and more root area. Now consider what happens to the roots and shoots when the growing season is over, when the leaves fall and the plants die. The new-grown biomass decays and is eaten by fungi or microbes. Some of it returns to the atmosphere, and some of it is converted into topsoil. On the average, more of the above-ground growth will return to the atmosphere, and more of the below-ground growth will become topsoil. So the plants with increased root-to-shoot ratio will cause an increased transfer of carbon from the atmosphere into topsoil. If the increase in atmospheric carbon dioxide due to fossil-fuel burning has caused an increase in the average root-to-shoot ratio of plants over large areas, then the possible effect on the topsoil reservoir will not be small. At present we have no way to measure or even to guess the size of this effect. The aggregate biomass of the topsoil of the planet is not a measurable quantity. But the fact that the topsoil is unmeasurable does not mean that it is unimportant.

At present we do not know whether the topsoil of the United States is increasing or decreasing. Over the rest of the world, because of large-scale deforestation and erosion, the topsoil reservoir is probably decreasing. We do not know whether intelligent land management could increase the growth of the topsoil reservoir by four billion tons of carbon per year, the amount needed to stop the increase of carbon dioxide in the atmosphere. All that we can say for sure is that this is a theoretical possibility and ought to be seriously explored.

Oceans and Ice Ages

Another problem that has to be taken seriously is a slow rise of sea level, which could become catastrophic if it continues to accelerate. We have accurate measurements of sea level going back two hundred years. We observe a steady rise from 1800 to the present, with an acceleration during the last fifty years. It is widely believed that the recent acceleration is due to human activities, since it coincides in time with the rapid increase of carbon dioxide in the atmosphere. But the rise from 1800 to 1900 was probably not due to human activities. The scale of industrial activities in the nineteenth century was not large enough to have had measurable global effects. So a large part of the observed rise in sea level must have other causes. One possible cause is a slow readjustment of the shape of the earth to the disappearance of the northern ice sheets at the end of the ice age twelve thousand years ago. Another possible cause is the large-scale melting of glaciers, which also began long before human influences on climate became significant. Once again, we have an environmental danger whose magnitude cannot be predicted until we know more about its causes (Munk 2002).

The most alarming possible cause of sea-level rise is a rapid disintegration of the West Antarctic ice sheet, which is the part of Antarctica where the bottom of the ice is far below sea level. Warming seas around the edge of Antarctica might erode the ice cap from below and cause it to collapse into the ocean. If the whole of West Antarctica disintegrated rapidly, sea level would rise by five meters, with disastrous effects on billions of people. However, recent measurements of the ice cap show that it is not losing volume fast enough to make a significant contribution to the currently observed sea-level rise. It appears that the warming seas around Antarctica are causing an increase in snowfall over the ice cap, and the increased snowfall on top roughly cancels out the decrease of ice volume caused by erosion at the edges. The same changes, increased melting of ice at the edges and increased snowfall adding ice on top, are also observed in Greenland. In addition, there is an increase in snowfall over the East Antarctic ice cap, which is much larger and colder and is in no danger of melting. This is another situation where we do not know how much of the environmental change is due to human activities and how much to long-term natural processes over which we have no control.

Another environmental danger that is even more poorly understood is the possible coming of a new ice age. A new ice age would mean the burial of half of North America and half of Europe under massive ice sheets. We know that there is a natural cycle that has been operating for the last eight hundred thousand years. The length of the cycle is a hundred thousand years. In each hundred-thousand-year period, there is an ice age that lasts about ninety thousand years and a warm interglacial period that lasts about ten thousand years. We are at present in a warm period that began twelve thousand years ago, so the onset of the next ice age is overdue. If human activities were not disturbing the climate, a new ice age might already have begun. We do not know how to answer the most important question: Do our human activities in general, and our burning of fossil fuels in particular, make the onset of the next ice age more likely or less likely?

There are good arguments on both sides of this question. On the one side, we know that the level of carbon dioxide in the atmosphere was much lower during past ice ages than during warm periods, so it is reasonable to expect that an artificially high level of carbon dioxide might stop an ice age from beginning. On the other side, the oceanographer Wallace Broecker (1997) has argued that the present warm climate in Europe depends on a circulation of ocean water, with the Gulf Stream flowing north on the surface and bringing warmth to Europe, and a countercurrent of cold water flowing south in the deep ocean. So a new ice age could begin whenever the cold deep countercurrent is interrupted. The countercurrent could be interrupted when the surface water in the Arctic becomes less salty and fails to sink, and the water could become less salty when the warming climate increases the Arctic rainfall. Thus Broecker argues that a warm climate in the Arctic may paradoxically cause an ice age to begin. Since we are confronted with two plausible arguments leading to opposite conclusions, the only rational response is to admit our ignorance. Until the causes of ice ages are understood, we cannot know whether the increase of carbon dioxide in the atmosphere is increasing or decreasing the danger.

The Wet Sahara

My second heresy is also concerned with climate change. It is about the mystery of the wet Sahara. This is a mystery that has always fascinated me. At many places in the Sahara desert that are now dry and unpopulated, we find rock paintings showing people with herds of animals. The paintings are abundant, and some of them are of high artistic quality, comparable with the more famous cave paintings in France and Spain. The Sahara paintings are more recent than the cave paintings. They come in a variety of styles and were probably painted over a period of several thousand years. The latest of them show Egyptian influences and may be contemporaneous with early Egyptian tomb paintings. Henri Lhote's book *The Search for the Tassili Frescoes* (1958), is illustrated with reproductions of fifty of the paintings. The best of the herd paintings date from roughly six thousand years ago. They are strong evidence that the Sahara at that time was wet. There was enough rain to support herds of cows and giraffes, which must have grazed on grass and trees. There were also some hippopotamuses and elephants. The Sahara then must have been like the Serengeti today.

At the same time, roughly six thousand years ago, there were deciduous forests in northern Europe where the trees are now conifers, proving that the climate in the far north was milder than it is today. There were also trees standing in mountain valleys in Switzerland that are now filled with famous glaciers. The glaciers that are now shrinking were much smaller six thousand years ago than they are today. Six thousand years ago seems to have been the warmest and wettest period of the interglacial era that began twelve thousand years ago when the last ice age ended. I would like to ask two questions. First, if the increase of carbon dioxide in the atmosphere is allowed to continue, shall we arrive at a climate similar to the climate of six thousand years ago when the Sahara was wet? Second, if we could choose between the climate of today with a dry Sahara and the climate of six thousand years ago with a wet Sahara, should we prefer the climate of today? My second heresy answers yes to the first question and no to the second. It says that the warm climate of six thousand years ago with the wet Sahara is to be preferred, and that increasing carbon dioxide in the atmosphere may help to bring it back. I am not saying that this heresy is true. I am only saying that it will not do us any harm to think about it.

The biosphere is the most complicated of all the things we humans have to deal with. The science of planetary ecology is still young and undeveloped. It is not surprising that honest and well-informed experts can disagree about facts. But beyond the disagreement about facts, there is another, deeper disagreement about values. The disagreement about values may be described in an oversimplified way as a disagreement between naturalists and humanists. Naturalists believe that nature knows best. For them the highest value is to respect the natural order of things. Any gross human disruption of the natural environment is evil. Excessive burning of fossil fuels is evil. Changing nature's desert, either the Sahara desert or the ocean desert, into a managed ecosystem where giraffes or tuna fish may flourish, is likewise evil. Nature knows best, and anything we do to improve upon Nature will only bring trouble.

The humanist ethic begins with the belief that humans are an essential part of nature. Through human minds the biosphere has acquired the capacity to steer its own evolution, and now we are in charge. Humans have the right and the duty to reconstruct nature so that humans and biosphere can both survive and prosper. For humanists, the highest value is harmonious coexistence between humans and nature. The greatest evils are poverty, underdevelopment, unemployment, disease, and hunger, all the conditions that deprive people of opportunities and limit their freedoms. The humanist ethic accepts an increase of carbon dioxide in the atmosphere as a small price to pay if worldwide industrial development can alleviate the miseries of the poorer half of humanity. The humanist ethic accepts our responsibility to guide the evolution of the planet.

The sharpest conflict between naturalist and humanist ethics arises in the regulation of genetic engineering. The naturalist ethic condemns genetically modified food crops and all other genetic engineering projects that might upset the natural ecology. The humanist ethic looks forward to a time not far distant when genetically engineered food crops and energy crops will bring wealth to poor people in tropical countries, and incidentally give us tools to control the growth of carbon dioxide in the atmosphere. Here I must confess my own bias. Since I was born and brought up in England, I spent my formative years in a land with great beauty and a rich ecology which is almost entirely man-made. The natural ecology of England was uninterrupted and rather boring forest. Humans replaced the forest with an artificial landscape of grassland and moorland, fields and farms, with a much richer variety of plant and animal species. Quite recently, only about a thousand years ago, we introduced rabbits, a nonnative species that had a profound effect on the ecology. Rabbits opened glades in the forest where flowering plants now flourish. There is no wilderness in England, and yet there is plenty of room for wildflowers and birds and butterflies as well as a high density of humans. Perhaps that is why I am a humanist.

Taking Turns at Being Top

To conclude this chapter I come to my third and last heresy. My third heresy says that the United States has less than a century left of its turn as top nation. Since the modern nation-state was invented around the year 1500, a succession of countries have taken turns at being top nation, first Spain, then France, Britain, America. Each turn lasted about 150 years. The United States' turn began in 1920, so it should end about 2070. The reason why each top nation's turn comes to an end is that the top nation becomes overextended, militarily, economically, and politically. Greater and greater efforts are required to maintain the number one position. Finally the overextension becomes so extreme that the structure collapses. Already we can see in the American posture today some clear symptoms of overextension. Who will be the next top nation? China is the obvious candidate. After that it might be India or Brazil. We should be asking ourselves, not how to live in an America-dominated world, but how to prepare for a world that is not America-dominated. That may be the most important problem for the next generation of Americans to solve. How does a people that thinks of itself as number one yield gracefully to become number two?

I am telling the next generation of young students, who will still be alive in the second half of our century, that misfortunes are on the way. Their precious Ph.D., or whichever degree they went through long years of hard work to acquire, may be worth less than they think. Their specialized training may become obsolete. They may find themselves overqualified for the available jobs. They may be declared redundant. The country and the culture to which they belong may move far away from the mainstream. But these misfortunes are also opportunities. It is always open to them to join the heretics and find another way to make a living. With or without a Ph.D., there are big and important problems for them to solve.

I will not attempt to summarize the lessons that my readers should learn from these heresies. The main lesson that I would like them to take home is that the long-range future is not predetermined. The future is in their hands. The rules of the world-historical game change from decade to decade in unpredictable ways. All our fashionable worries and all our prevailing dogmas will probably be obsolete in fifty years. My heresies will probably also be obsolete. It is up to them to find new heresies to guide our way to a more hopeful future.

Bad Advice to a Young Scientist

Sixty years ago, when I was a young and arrogant physicist, I tried to predict the future of physics and biology. My prediction was an extreme example of wrongness, perhaps a world record in the category of wrong predictions. I was giving advice about future employment to Francis Crick, the great biologist who died in 2005 after a long and brilliant career. He discovered, with Jim Watson, the double helix. They discovered the double helix structure of DNA in 1953 and thereby gave birth to the new science of molecular genetics. Eight years before that, in 1945, before World War II came to an end, I met Francis Crick for the first time. He was in Fanum House, a dismal office building in London where the Royal Navy kept a staff of scientists. Crick had been working for the Royal Navy for a long time and was depressed and discouraged. He said he had missed his chance of ever amounting to anything as a scientist. Before World War II, he had started a promising career as a physicist. But then the war hit him at the worst time, putting a stop to his work in physics and keeping him away from science for six years. The six best years of his life, squandered on naval intelligence, lost and gone forever. Crick was good at naval intelligence and did important work for the navy. But military intelligence bears the same relation to intelligence as military music bears to music. After six years doing this kind of intelligence, it was far too late for Crick to start all over again as a student and relearn all the stuff he had forgotten. No wonder he was depressed. I came away from Fanum House thinking, “How sad. Such a bright chap. If it hadn't been for the war, he would probably have been quite a good scientist."

A year later, I met Crick again. The war was over and he was much more cheerful. He said he was thinking of giving up physics and making a completely fresh start as a biologist. He said the most exciting science for the next twenty years would be in biology and not in physics. I was then twenty-two years old and very sure of myself. I said, "No, you're wrong. In the long run biology will be more exciting, but not yet. The next twenty years will still belong to physics. If you switch to biology now, you will be too old to do the exciting stuff when biology finally takes off." Fortunately, he didn't listen to me. He went to Cambridge and began thinking about DNA. It took him only seven years to prove me wrong. The moral of this story is clear. Even a smart twenty-two-year-old is not a reliable guide to the future of science. And the twenty-two-year-old has become even less reliable now that he is eighty-two.

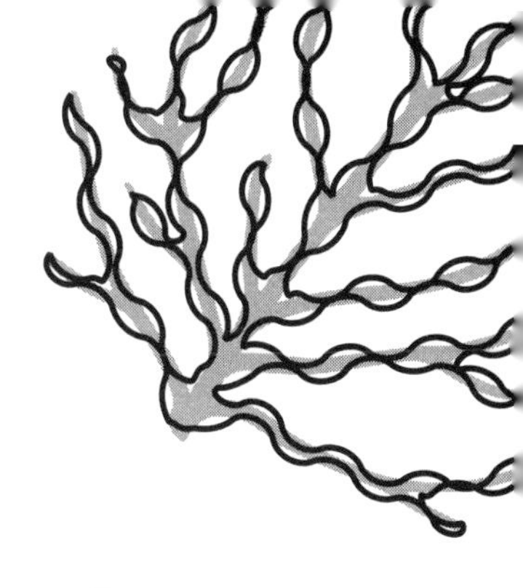

4
A Friendly Universe

Order and Disorder

I look at the universe from the point of view of a physicist and ask how it happens that the laws of physics have made our universe hospitable to life. I will be painting a picture of the universe with a very broad brush, not looking at details but at the general scheme of things. It turns out that our existence as living creatures depends on general features of the universe such as the laws of gravitation and the prevalence of phase transitions. I will try to explain in nontechnical language how these features of the universe work. I find it remarkable that we can start with a few general principles and end with a pretty good understanding of why the universe is friendly.

In my broad-brush picture, the evolution of the universe is dominated by the paradox of order and disorder. The paradox is the apparent contradiction between two facts. On the one hand, the total disorder in the universe, as measured by the quantity that physicists call entropy, increases steadily as we go from past to future. On the other hand, the total order in the universe, as measured by the complexity and permanence of organized structures, also increases steadily as we go from past to future. How can it happen that both order and disorder are constantly increasing with time? This is the paradox that we have to understand. The paradox exists, whether or not the universe contains life. It is connected with life, because life is a conspicuous example of a process involving an intimate mixture of order and disorder. If we can explain the paradox, this may help us to understand life. But I talk about the paradox from the point of view of a physicist. It is for the biologists to say whether it is relevant to the processes of life.

Let me first tell you how this chapter is organized. The first half of the chapter will be describing how things happen, the second half will be analyzing why things happen. The first half is concerned with observational facts, the second half with theoretical explanations. I am trying to reconcile the theoretical law of increasing disorder in the universe with the evidence for increasing order in the universe as we observe it. I will discuss four explanations of the paradox, four reasons why order can grow against the flow of increasing entropy. The first two explanations are descriptive, the third and fourth are analytical. As usually happens when we meet a fundamental paradox in science, we can only understand it by looking at it from several points of view. The paradox is too subtle to be pinned down by a single explanation. The descriptive part of the chapter describes two general causes of increasing order in the universe. These are phase transitions and symbiosis. We usually consider phase transitions to belong to physics and chemistry, symbiosis to biology. When I talk about symbiosis I will be stepping across the border into the territory of the biologists. I will explain why symbiosis plays as fundamental a role in astronomy as it does in biology. The analytical part of the chapter discusses the third and fourth reasons why order in the universe increases. These are the expansion of the universe and the peculiar thermodynamics of objects held together by gravity. These two explanations go more deeply into the nature of the paradox. In particular, the peculiar relation between thermodynamics and gravity is the most basic reason why our universe is hospitable to the evolution of life.

Phase Transitions

I now begin the descriptive part of the chapter. For readers who would like a more detailed picture, Eric Chaisson has written a good textbook describing the processes of cosmic evolution (Chaisson 1988). The simplest way in which order can arise out of disorder as the universe evolves is by the process of phase transition. A phase transition is an abrupt change in the physical or chemical properties of matter, usually caused by heating or cooling. Familiar examples of phase transitions are the freezing and boiling of water, the magnetization and demagnetization of iron, the precipitation of rain and snow from water vapor dissolved in air. In many of these transitions, the warmer phase is a uniform disordered mixture while the cooler phase divides itself into two separate components with a more ordered structure. Such transitions are called order-disorder transitions. The transition from warm humid air to cold dry air plus snowflakes is a spectacular example of an order-disorder transition. The snowflakes appear with a beautiful crystalline structure that was totally absent from the humid air out of which they arose. Also, by the action of the earth's gravity, the snowflakes spontaneously separate themselves from the air and fall to the ground. At all stages in the evolution of the universe we see order-disorder transitions with the same two characteristic features, first the sudden appearance of structures that did not exist before, and second the physical separation of the newborn structures into different regions of space.

Another name for the process of phase transition from disorder to order is symmetry breaking. From a mathematical point of view, a disordered phase has a higher degree of symmetry than an ordered phase. For example, the environment of a molecule of water dissolved in humid air is the same in all directions, while the environment of the same molecule after it is precipitated into a snowflake is a regular crystal with crystalline axes oriented along particular directions. The molecule sees its environment change from the greater symmetry of a sphere to the lesser symmetry of a hexagonal prism. The change in the environment from disorder to order is associated with a sudden loss of symmetry. Sudden loss of symmetry is characteristic of many of the most important phase transitions in the history of the universe.

In the earliest stages of its history, the universe was hot and dense and rapidly expanding. The matter and radiation were then totally disordered and uniformly mixed. One of the greatest of all symmetry breakings was the separation of the universe into two phases, one phase containing most of the matter and destined to condense later into galaxies and stars, the other phase containing most of the radiation and destined to become the intergalactic void. The separation happened as soon as the universe became transparent enough so that large lumps of matter pulled together by their own gravitation could radiate away their gravitational energy into the surrounding void. As a result of this transition, the universe lost its original spatial symmetry. Before the transition, it had the symmetry of a uniform space. After the transition, it became a collection of lumps with no large-scale symmetry. The same process of symmetry breaking was then repeated successively on smaller and smaller scales.

The first generation of lumps was given the name *bliny,* Russian for "pancakes," by the astrophysicist Yakov Zeldovich. Zeldovich imagined the universe after the first phase separation to have the structure of a foam composed of gigantic bubbles (Silk, Szalay, and Zeldovich 1983). The interiors of the bubbles are the voids containing mostly radiation and little matter. The flat walls between the bubbles are the pancakes containing most of the matter. Recent observations of the large-scale distribution of matter in the universe confirm that Zeldovich's picture is close to the truth. The division of the universe into pancakes and voids is a result of the physical separation of matter and radiation. After the universe had cooled and expanded for a while longer, the matter in the pancakes again separated into two phases, one cooler and composed of neutral atoms, one hotter and composed of ionized plasma. The cool phase condensed into galaxies, the second-generation lumps, while the hot phase filled the space between galaxies. The third phase separation happened when the atoms in galaxies became cool enough to combine into molecules. The gas then divided itself into a cool molecular phase and a warm atomic phase. The molecular phase condensed into giant molecular clouds, the third-generation lumps, while the atomic phase occupied the rest of the galaxies. The fourth phase separation happened when the molecular gas became cool enough to precipitate solid particles, grains of ice and dust mixed with the gas. The gas then separated into a dust-rich phase condensing into stars and planets, the fourth-generation lumps, and a dust-poor phase recycling matter into the interstellar gas. At each stage of the evolution, matter became differentiated into new phases, and the structure of the universe became more organized and more finely tuned.

The processes of differentiation and fine-tuning did not stop after the stars and planets were formed. After the earth had condensed out of the interstellar dust, a new world of opportunities opened for separation of phases and growth of structures. First came the separation of the interior of the earth into its main components: core, mantle, and crust. This separation is an active process that is probably still continuing. Next came the separation of the earth's surface into land, ocean, and atmosphere. This is also a continuing process, with water constantly circulating from the ocean into the atmosphere, onto the land, and back to the ocean. By the action of water, the crust of the earth became chemically differentiated, with rich concentrations of rare elements in some places, huge masses of common salt in others. The third great process transforming the earth is the division of the crust into plates with formation and destruction of crust at plate boundaries. This process is known as plate tectonics. The plate boundaries are of two types, rising mid-ocean ridges and sinking ocean trenches. At the ridges, new crust is produced by upwelling of liquid rock from the mantle below. At the trenches, old crust is destroyed by sinking back into the mantle. The action of plate tectonics ensures that the geography of the earth is constantly changing. New continents such as Australia and Antarctica are formed by splitting up old ones. New chains of mountains such as the Himalaya rise where the formerly separate continents India and Asia collide. And new chains of islands such as Japan and the Aleutians are created by volcanos rising from the edge of a descending plate. Plate tectonics is a powerful force constantly giving the earth new structures. But it is not the most powerful of the forces at work on our planet. The fourth process creating structure and order on the earth is the most powerful of all. The fourth process, after the separation of core and mantle and crust, after the separation of land and ocean and atmosphere, after the shaping of continents by plate tectonics, is life. Life appeared here between three and four billion years ago and transformed the planet more profoundly than all the other processes together.

The transition from dead to living was a phase transition of a new type. It was a transition from disorder to order in which the ordered phase acquired the ability to perpetuate itself after the conditions that caused it to appear had changed. There are many theories of the origin of life, and there is no direct evidence to decide which theory is true. All that we know for sure is that a complicated mixture of organic chemicals made the transition to an ordered phase that could grow and reproduce itself and feed on its surroundings. And then, after the ordered phase was once established, it possessed the flexibility to mutate and evolve into a million different forms, and the different forms competed with one another to gain the advantages of being more prolific and more rugged than their competitors. Out of their mutation and competition emerged the grand panorama of structures that we see around us today, the birds and flowers and trees and viruses and whales and bees, the coral reefs and rain forests and peat bogs, the chalk in our hills and the free oxygen in our atmosphere. Life has given to our planet a richness of structure that we see nowhere else in the universe.

After the great phase transition from nonliving to living, the evolution of life has proceeded through a succession of lesser phase transitions, from microbes to animals and plants, from asexual to sexual, from marine to terrestrial, each transition giving rise to an explosion of new structures upon which later transitions can build. The diversification of new forms of life on the earth is in many respects similar to the diversification of new types of celestial objects—galaxies and dust clouds and stars and planets—in the universe as it was before life appeared. The evolution of life may be seen as a continuation of the evolution of the universe. Both in the nonliving universe and on the living earth, evolution alternates between long periods of stability and short periods of rapid change. During the periods of rapid change, old structures become unstable and divide into new structures. During the periods of stability, the new structures are consolidated and fine-tuned while the environment to which they are adapted seems eternal. Then the environment crosses some threshold that plunges the existing structures into a new instability, and the cycle of evolution starts again.

Symbiosis

Phase transitions are one of the two main driving forces of evolution, creating order and structure both in the universe and in the biosphere. The other main driving force is symbiosis. Symbiosis is the reattachment of two structures, after they have been detached from each other and have evolved along separate paths for a long time, so as to form a combined structure with behavior not seen in the separate components. Symbiosis is a familiar concept both in biology and in astronomy. In biology, almost all higher plants and animals make use of symbiotic bacteria to perform many of their metabolic functions. Nitrogen-fixing bacteria in the roots of soybean plants and cellulose-digesting bacteria in the stomachs of cows are two well-known examples.

Symbiosis played an even more fundamental role in the evolution of modern cells from primitive cells. Modern cells contain inside them little structures called mitochondria (which enable animals to derive energy from food and oxygen) and chloroplasts (which enable plants to derive energy from sunlight). The mitochondria and chloroplasts that are essential components of modern cells were once independent free-living creatures. We know this because they still carry with them their own little DNA genomes, separate from the genome of the host cell. They first invaded the ancestral primitive cell from the outside and then became adapted to living inside. The combined cell then learned to coordinate the activities of its component parts, so that it acquired a complexity of structure and function that neither component could have evolved separately. In this way symbiosis allows evolution to proceed in giant steps. A symbiotic creature can jump from simple to complicated structures much more rapidly than a creature evolving by the normal processes of genetic mutation (Margulis 1981). For readers who are interested in a more detailed account of the role of biological symbiosis in the evolution of higher organisms, I recommend a delightful essay by Lynn Margulis and Michael Dolan (1997). The essay is concerned with the centriole, the apparatus that organizes the process of cell division in modern cells, pushing the chromosomes around and arranging them neatly so that the two daughter cells inherit exactly equal shares of the genome. Margulis had long conjectured that the centriole was an even more ancient invader that had moved into the cells from outside before the ancestors of mitochondria and chloroplasts. To prove this conjecture, she needed to find a creature that preserves a relic of the ancient centriole DNA outside its own genome, just as we humans preserve mitochondrial DNA in our cells outside our genomes. She and Dolan found the creature, a parasitic microbe called Calonympha that lives in the guts of termites. Calonympha was the answer to Margulis's question. It has abundant centrioles with their own packages of DNA, and it has not a single mitochondrion.

Since I am a physicist and not a biologist, I know more about symbiosis in the nonliving universe. Symbiosis is as frequent in the sky as it is in biology. Astronomers are accustomed to talking about symbiotic stars. The basic reason why symbiosis is important in astronomy is the double mode of action of gravitational forces. When gravity acts upon a uniform distribution of matter occupying a large volume of space, the first effect of gravity is to concentrate the matter into lumps separated by voids. The separated lumps then differentiate and evolve separately along different evolutionary histories. They become distinct types of object. But then, after a period of separate existence, gravity acts in a second way to bring lumps together and bind them into pairs. The binding into pairs is a sporadic process depending on chance encounters. It usually takes a long time for a lump to be bound into a pair. But the universe has plenty of time. After a few billion years, a large fraction of objects of all sizes become bound in symbiotic systems, either in pairs or in clusters. Once they are bound together by gravity, dissipative processes of various kinds tend to bring them closer together. As they come closer together, they interact with one another more strongly and the effects of symbiosis become more striking.

Examples of astronomical evolution caused by symbiosis are to be seen wherever one looks in the sky. On the largest scale, symbiotic pairs and clusters of galaxies are very common. When galaxies come into close contact, their internal evolution is often profoundly modified. A common sign of symbiotic activity is an active galactic nucleus. An active nucleus is seen in the sky as an intensely bright source of light at the center of a galaxy. The varieties of active galactic nuclei commonly arise from the symbiotic effects of other galaxies nearby. The probable cause of the intense light is gas falling into a black hole at the center of one galaxy as a result of the gravitational perturbations caused by another galaxy. It happens frequently that big galaxies swallow small galaxies. Nuclei of swallowed galaxies are observed inside the swallower, like mouse bones in the stomach of a snake. This form of symbiosis is known as galactic cannibalism.

On the scale of stars, we can distinguish even more different types of symbiosis, because there are many types of star and many stages of evolution for each of the stars in a symbiotic pair. The most spectacular symbiotic pairs have one component that is highly condensed, a white dwarf or a neutron star or a black hole, and the other component a normal star. If the two stars are orbiting around each other at a small distance, gas will spill over from the normal star into the deep gravitational field of the condensed star. The gas falling into the deep gravitational field will become intensely hot and will produce a variety of unusual effects such as recurrent nova outbursts, intense bursts of X-rays, and rapidly flickering light variations. The more common and less spectacular symbiotic pairs consist of two normal stars orbiting around each other close enough so that mass is exchanged between them. These pairs are seen as optically variable stars, often surrounded by rings of gas and dust. They are important to the long-term evolution of our galaxy because they recycle matter from the stars back into the interstellar gas and dust out of which new generations of stars will be formed.

Finally, the least common and least conspicuous type of symbiotic pair consists of two condensed stars. These can be seen with radio telescopes if one component of the pair is a pulsar, a neutron star emitting radio pulses as it rotates. One such pair, a symbiosis of two neutron stars, was discovered by the radio astronomers Joseph Taylor and Russell Hulse, who received the Nobel Prize in 1993 for the discovery. This symbiotic pair of neutron stars is scientifically important because it gave us the first clear evidence for the existence of gravitational waves. An even more informative symbiosis of two neutron stars was recently discovered by radio astronomers in Australia. In the Australian neutron-star pair, both components are pulsars, and the dynamics of the system can be studied in exquisite detail. Another reason why symbiotic neutron stars are important is that the drag produced by gravitational waves brings them steadily closer together as time goes on. Ultimately they are brought so close together that they become dynamically unstable and fall together into a single star with a titanic splash of spiral arms carrying away their angular momentum. The process of collapse takes only a few thousands of a second and must result in a huge burst of outgoing gravitational waves. This collapse of symbiotic neutron stars might also explain some of the mysterious bursts of gamma rays that are seen coming from random directions in the sky at a rate of about one per day. Gamma-ray bursts are the most violent events in the whole universe, even more violent than the supernova explosions that occur when neutron stars are born. Symbiosis is not a minor factor in the evolution of the universe. It is one of the main driving forces.

But this book is supposed to be about the universe as a home for life, and I must not digress further into the technical problems of astronomy. Let me come back to the subject of life. From the point of view of life, the most important example of astronomical symbiosis is the symbiosis of the earth and the sun. The whole system of sun and planets and satellites, the system that we call the solar system, is a typical example of astronomical symbiosis. At the beginning, when the solar system was formed, the sun and the earth were born with totally different chemical compositions and physical properties. The sun was made mainly of hydrogen and helium, the earth was made of heavier elements such as oxygen and silicon and iron. The sun was physically simple, a sphere of gas heated at the center by the burning of hydrogen into helium and shining steadily for billions of years. The earth was physically complicated, partly liquid and partly solid, its surface frequently transformed by phase transitions of many kinds. The symbiosis of these two contrasting worlds made life possible. The earth provided chemical and environmental diversity for life to explore. The sun provided physical stability, a steady input of energy on which life could rely. The combination of the earth's variability with the sun's constancy provided the conditions in which life could evolve and prosper.

In addition to the sun and the planets and their satellites, the solar system also contains a large number of asteroids and comets, smaller objects gravitationally bound to the sun but not sharing in the orderly motions of the planets. The asteroids and comets are an important part of the symbiosis that binds the system together. Since they have disordered motions, they occasionally collide with planets and produce catastrophic disturbances of the local environment. Traces of these impacts are visible on the surface of the earth, and even more visible on the moon. Impacts large enough to affect the whole earth and cause extinctions of species on a global scale occur about once in a hundred million years. The random obliteration of ecologies by major impacts has been a part of the history of life on Earth since the beginning. It is likely that these catastrophes drove evolution forward by destroying species that were too well adapted to static environments, making room for species that were adaptable to harsher and more rapidly changing conditions. Without the occasional impact catastrophe to reward adaptability, it is unlikely that our own species would have emerged. We are among the most adaptable of species, offspring of a symbiosis in which sun, planets, asteroids, and comets all played an essential part.

The Heat Death

That is all I have to say about symbiosis. The next subject I want to talk about is entropy. Here I am obliged to introduce a little bit of technical jargon. I need to define what a physicist means by disorder or entropy. When we say that a physical object is disordered, we mean that it might be in any one of a big number of possible states, and we have no way to tell which particular state it is in. The entropy measures our degree of ignorance of the precise state of the object. It is the amount of information we would need in order to know which of the possible states the object is in. Since we are accustomed to code information in computers with binary arithmetic, using a sequence of zeros and ones to carry a message, the practical unit of information is a binary digit or bit. The bit is a very small unit. It usually takes a few bits to describe the state of an atom, so the entropy of you or me is a few times the number of atoms that we contain. That is a huge number. Each of us contains a huge amount of entropy, which is as it should be, because we are mostly made up of water and we have no idea what all those water molecules are doing. Readers who do not like scientific jargon do not need to bother with the exact definition of entropy. They can simply say that the entropy of an object is the amount of disorder that it contains. This last statement is a good enough definition of entropy for politicians or poets, or for anyone who is not interested in mathematical precision.

Entropy is a precise measure of disorder. It is important because there is a fundamental physical law called the second law of thermodynamics. The second law says that in any closed system, which means any collection of objects isolated from contact with the world outside, the entropy can never decrease with time. In any closed system, as time goes on, entropy either stays constant or increases. If you consider the whole universe as the system, there is nothing outside to make contact with it, and so the second law applies to it. The entropy of the universe can never decrease. In fact, the entropy of the universe is constantly increasing. This is the first half of the paradox that I mentioned earlier. We live in a universe of constantly increasing disorder.

The science of thermodynamics describes the way energy and entropy move around in the universe. Every object carries a certain amount of energy and a certain amount of entropy. The thermodynamic temperature of the object is defined to be the amount of energy we have to add in order to increase the entropy by one unit. The higher the temperature, the more energy we have to add per unit of entropy. The lower the temperature, the more entropy we have to add per unit of energy. Suppose that we have two objects at different temperatures. Suppose, if possible, that some energy moves from the colder object to the hotter object. Then the colder object must lose more entropy than the hotter object gains. This means that the total entropy of the two objects together must decrease. The second law of thermodynamics says that this is impossible, so long as the two objects are isolated from their surroundings. The practical effect of the second law is that energy can move only in one direction, from hotter to colder, and not from colder to hotter, unless it is driven by some outside agency such as a refrigerator. Inside any closed system of objects, the flow of energy always runs from hotter to colder, which is the direction of increasing total entropy.

When the second law was first discovered in the nineteenth century, many people considered it to be a disaster for humanity and for life in general. They made two additional assumptions that looked reasonable but eventually turned out to be wrong. The two assumptions were (1) the universe considered as a whole is a stationary system, and (2) all objects in the universe have positive specific heat. The specific heat of an object is the amount of energy you must add to the object to produce a unit rise in temperature. To say that an object has a positive specific heat means that its total content of energy increases as its temperature increases. If you add energy to the object, its temperature goes up.

All the objects we are familiar with in daily life, solids and liquids and gases, pots and pans, soup and fish and eggs, certainly behave in this way. The art of cooking depends on it. When you are cooking, you add energy to an object in order to increase its temperature. It would be a strange kind of cooking if the fish became colder when you put it over the fire. Common sense, based on thousands of years of experience of ordinary people using fire for heating and cooking, made it seem obvious that everything in the universe should have positive specific heat. I will explain later how it happens that this common sense can sometimes be wrong.

If we assume, as the experts in the nineteenth century assumed, that the universe is stationary and that all specific heats are positive, then we reach a dismal conclusion. The second law of thermodynamics says that energy can only flow from hotter to colder objects. If all specific heats are positive, this means that the flow of energy from one object to another always causes the hotter object to become cooler and the cooler object to become hotter. In this case, all differences of temperature between different objects become smaller and smaller as time goes on. If we wait long enough, all differences of temperature will tend to disappear and the whole universe will tend to a state of uniform temperature. The state of uniform temperature is also the state of maximum entropy or maximum disorder. When the universe has reached the state of uniform temperature we say that it is in thermodynamical equilibrium. Once the universe is in thermodynamical equilibrium, no energy can flow from one object to another. The temperature of every object remains constant, and all thermodynamic processes stop. The nineteenth-century scientists who believed that all specific heats must be positive called this equilibrium state of the universe the "heat death." The heat death is the state of maximum entropy, a state that brings all physical change to a halt. Since life involves flows of energy between living creatures and their environment, life must disappear as the universe approaches the heat death. The nineteenth-century scientists deduced that life is doomed to ultimate extinction. Life is a manifestation of order; therefore life cannot survive when the universe inexorably slides toward the state of maximum disorder.

If the universe were really tending inexorably toward the heat death, then not only the existence of life but the existence of any kind of order would be paradoxical. In the long run, all kinds of order would be doomed to dissipate and disappear as the universe tends toward the state of maximum entropy. But when we look at the universe as it actually is, we see order and organization everywhere. The local corner of the universe where we live, with the nine planets revolving peacefully around the sun in almost circular orbits, already impressed our ancestors as a manifestation of the divine order of nature. On our own planet, the regular sequences of days and nights, summer and winter, birth and death, are also symbols of order. The Greek word *cosmos,* which we use as a synonym for "universe," actually means "order." And when we look deeper into the history of the universe, we see that order not only exists everywhere but is increasing as time goes on. Before the ordered system of the sun and the planets was formed, the sun and planets were a cloud of interstellar gas and dust with a far less ordered structure. Before the intricate ordered patterns of life, with trees and butterflies and birds and humans, grew to cover our planet, the earth's surface was a boring unstructured landscape of rock and sand. And before the grand ordered structures of galaxies and stars existed, the universe was a rather uniform and disordered collection of atoms. What we see in the real universe is the opposite of a heat death. We see the universe growing visibly more ordered and more lively as it grows older.

The final two sections of this chapter explain two reasons why the universe never runs into a heat death: the expansion of the universe and the peculiar thermodynamic behavior of gravity. These two reasons give us at the same time a solution of the order-disorder paradox. The old argument for the heat death was based on two assumptions, that the universe was stationary and that all specific heats were positive. Both assumptions are wrong, the first because the universe is expanding and the second because objects held together by gravity often have negative specific heat.

The Expanding Universe

The fact that the universe is expanding rather than stationary was discovered by Edwin Hubble in the 1920s. The discovery came as an unwelcome shock to most of the theoretical scientists of that time, and especially to Einstein. For reasons that are now difficult to understand, Einstein had a strong prejudice that the universe ought to be stationary. This prejudice did not arise from the theory of general relativity, which he had discovered a few years earlier. The theory of general relativity actually runs into difficulties if the universe is stationary. Einstein had decided to change his theory by adding an additional term, the famous "cosmological constant," which made the theory more complicated, just in order to allow the universe to be stationary. After Hubble's discovery, Einstein abandoned the cosmological constant and resigned himself to living in a nonstationary universe. But the old prejudice in favor of a stationary universe was still strong. It was shared by many other great scientists besides Einstein. It probably arose long ago from the ancient Greek view of the celestial sphere as a region of unchanging and perpetual peace. It was this prejudice that caused the scientists of the nineteenth century to take seriously the idea of a universal heat death. And the idea of the heat death remained fixed in many people's minds even after Hubble had made it unnecessary.

How does it happen that in an expanding universe we can have increasing order and increasing disorder at the same time? From an intuitive point of view, the answer to this question is almost obvious. The expansion gives more space for the physical separation of order and disorder. Order can increase in one part of the universe and disorder in another. The physical separation between order and disorder allows each to increase without coming into conflict. As a practical example to illustrate how this works, consider the planet Earth. Order increases on planet Earth when water and carbon dioxide and minerals are converted by the energy of sunlight into rice and fish and humans. To pay for this increase of order, disorder in the form of infrared heat radiation is radiated away from the earth into space. To satisfy the second law of thermodynamics, the amount of disorder in the outgoing heat radiation is always greater than the amount of order in the new growth here on Earth. But the disorder in the heat radiation does not disturb us because it never comes back to Earth. If we lived in a stationary nonexpanding universe, the whole universe would gradually become filled with heat radiation, and the radiation emitted from the earth would ultimately come back to haunt us. After a very long time, the sky would become filled with heat radiation at the same temperature as the earth, and we would be stifled by our own waste heat. The heat death would then be a reality. But because the volume of the universe is steadily expanding, the heat radiation becomes more and more dilute as time goes on, and its temperature decreases. The amount of disorder in the heat radiation is constantly growing, but its temperature and its density in space are diminishing so that it will never do us any harm.

The Peculiar Behavior of Gravity

The expanding universe solves the paradox of order and disorder for processes happening on the scale of the entire universe. The paradox on the universal scale is solved, because order can increase in one part of the universe while disorder increases in another part. But the paradox also exists on a local scale, for processes confined to local regions. For example, the formation of the sun and the earth and the other planets out of a condensing disk of interstellar gas was a local process. The condensation produced a system of orbiting objects with much more order than the gas cloud that gave them birth, while concentrating disorder into the interior of the sun. The same process of formation of the solar system could have happened in exactly the same way in a stationary universe or even in a contracting universe. The expansion of the universe had nothing to do with it. We therefore need another solution of the paradox, a solution that explains how order and disorder can both increase in a local neighborhood, independently of what the universe outside the neighborhood may be doing. We saw earlier, when I wrote about the heat death, that every local group of objects isolated from the rest of the universe must run into a heat death if the objects in the group have positive specific heat, that is to say, if each object gets hotter when you increase its energy. Therefore, to avoid the heat death locally, whether the universe is stationary or not, the group must contain one or more objects with negative specific heat. We escape from the paradox of order and disorder on the local level because, wherever we look in our universe, we find an abundance of objects with negative specific heat, objects that get hotter when they lose energy and get cooler when they gain energy. When you have even one object with negative specific heat, you can never have a heat death. If the object is hotter than its surroundings, energy will flow out of it and it will continue to get hotter, and if it is cooler than its surroundings, energy will flow into it and it will continue to get cooler.

Consider the example of the solar system. To solve the paradox we must find some object in the solar system that has a negative specific heat, an object that gets hotter as it loses energy. We certainly do not find such objects lying around on the earth. So far as we know, such objects also do not exist on the other planets. Only one possibility remains. The object with negative specific heat must be the sun. And indeed, when we examine the sun carefully, we find that its specific heat is in fact negative. The most obvious difference between the sun and the earth is the fact that the sun shines while the earth does not. Since the sun is shining, it is constantly losing energy. If the sun had a positive specific heat, it would become cooler as it loses energy, and it could not continue to shine. The mere fact that the sun continues to shine shows that its specific heat must be negative. When we calculate the evolution of the sun in detail, we find that it is gradually growing hotter and brighter as it radiates energy away. Since the sun was born four and a half billion years ago, its brightness has increased by about 30 percent. In the future it will continue to grow hotter, until after another five billion years it will evolve into a red giant star. It is the negative specific heat that allows the sun to continue shining for ten billion years and to provide a stable environment for the evolution of ordered structures upon the earth.

The negative specific heat of the sun is not an isolated accident. Not only the sun but the majority of luminous objects in the sky—stars, galaxies, gas clouds, and star clusters—also have negative specific heat. Their negative specific heat is not caused by some peculiarity of their structure or chemical composition. Negative specific heat is a consequence of the basic properties of gravity. The energy of coal and oil and nuclear fuels is positive, but the potential energy of gravitation is negative. For example, the gravitational energy of a rock on the surface of the earth is negative because it costs energy to push it off the earth into space. The gravitational energy of any massive object is negative because it costs energy to pull it apart. Gravity is unique among the forces of nature because its energy is always negative.

If any object is supported by a balance between the force of gravity pulling it together and the pressure force of hot gas pushing it apart, the specific heat of the object will be negative. Gravity and thermal gas pressure control the shape and size of the sun and of all ordinary stars. Objects with negative specific heat are abundant because gravitational energy is the dominant form of energy in the universe. Gravity and thermodynamics work together in a paradoxical way to produce objects with negative specific heat. This thermodynamic peculiarity of gravity is the last of my four solutions of the order-disorder paradox.

The negative gravitational energy that binds any collection of objects together becomes larger and more negative, the closer they come to one another. Also, their negative gravitational energy is twice as large as the positive energy of motion or pressure that holds them apart. The total energy of the system, gravity plus pressure, is then equal to minus the pressure energy. Consider again the case of the sun. If we imagine the sun to contract, its density and pressure and temperature will all increase, but its total energy, being predominantly gravitational and negative, will become more negative. An increase of temperature goes with a decrease of energy. Similarly, if the sun expands, a decrease of temperature will go with an increase of energy. The specific heat of the sun is negative because gravity dominates over gas pressure. For objects with negative specific heat, thermodynamic equilibrium is impossible. So long as the sun continues to shine, it will never cease to make possible the growth of order on the cooler celestial bodies orbiting around it.

The sources of the sun's energy are finite, and therefore it cannot continue shining forever. Five billion years from now, after it has reached its maximum brightness as a red giant, the sun will exhaust its reserves of nuclear energy and will become a white dwarf. It will continue to shine for a few hundred million years as a white dwarf, but it will then have a positive specific heat and will become cooler as it radiates away its remaining energy. Its final state will be a dark solid crystalline ball, mainly composed of carbon and oxygen, no longer shining and no longer immune to the heat death. It will slowly come into thermal equilibrium with its surroundings. In the end, it will approach a state of maximum entropy. It will then no longer be capable of importing disorder and exporting order, as it does now while it is still young and vigorous.

Stars heavier than the sun will end their lives differently. They will ultimately collapse either into neutron stars or into black holes. The heaviest stars will become black holes, while stars of intermediate mass will become neutron stars. Neutron stars behave thermodynamically like white dwarfs. At the end of their lives they have positive specific heat and cool quietly into a state of thermodynamical equilibrium. But black holes can never reach thermodynamical equilibrium and will always have negative specific heat. Black holes carry in their hidden depths a prodigiously large quantity of entropy. So long as black holes exist, they will provide a place where disorder can be imported and order exported. They give us a solution of the order-disorder paradox that will endure long after the sun and all the other visible objects in our sky have become cold and dark.

That is the end of my story. I do not have time to go more deeply into the beautiful theory of the thermodynamics of black holes, a theory created by the genius of Stephen Hawking. Black holes will be, in the long run, the dominating sources of energy and the dominating sinks of entropy in the universe. The properties of black holes will determine whether life can ultimately survive or not. The next chapter will give you a sketch of the distant future when life may be transforming and organizing the universe as a whole.

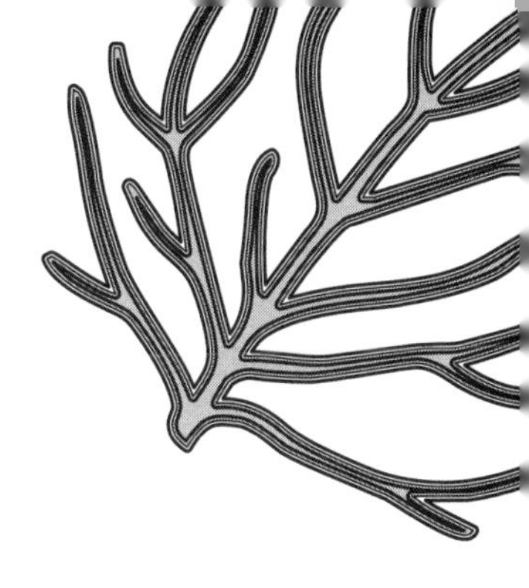

5
Can Life Go On Forever?

A Clash of Opinions

The question is whether life and intelligence can survive forever in an expanding universe that is constantly growing colder as time goes on. We cannot hope to answer this question with any certainty. We know very little about the nature of the universe and even less about the nature of life and intelligence. But we know enough to speculate about the problem, and that is what I will do in this chapter. I don't pretend to be able to predict the future. All I can do is explore the future, to see whether the survival of life is consistent with the laws of physics and the laws of information theory. If we find that the laws of physics and information theory make survival impossible, then we have answered the question in the negative. If we find that the laws do not forbid survival, we still do not know the answer. Survival might be impossible because of accidents of history or geography. We cannot in any case prove that life can survive. The best we can do is show that survival is not forbidden by the laws of nature as we know them today.

There are many uncertainties in our knowledge of the laws of nature. One of the most serious uncertainties is the question of proton decay. The proton is the nucleus of a hydrogen atom, and it is an essential component in the nuclei of all atoms. We have no evidence that the proton is unstable. All that we know for sure is that it lives for a very long time. Various so-called grand unified theories of elementary particles predict that protons should decay into lighter particles, mainly positrons and neutrinos, with a lifetime on the order of 10^{33} years. For readers who are not familiar with high powers of ten, I should explain that 10^{33} means a trillion times a trillion times a billion. It is, in short, an inordinately long time, but not too long to be measurable if protons are actually unstable. If the lifetime of a proton were 10^{33} years, it would mean that in a tank containing ten thousand tons of water three protons would decay each year on the average. With a great deal of care and a lot of heavy equipment, the decay of three protons per year would be detectable. Large underground detectors containing thousands of tons of water were built to observe the protons decaying, and no decay events were seen. The detectors turned out to be wonderful observatories for exploring the universe, since they detect neutrinos coming from the sky, but they have never detected a decaying proton. If the proton is unstable, the underground detectors tell us that its lifetime must be greater than 10^{33} years.

The majority of particle physicists still believe that protons are unstable. If this is true, all atoms are also unstable, and all matter will disintegrate into electrons and positrons within a finite time. The disappearance of ordinary matter will leave only an electron-positron plasma as a possible embodiment of life. If the density of the plasma decreases with time, it can last forever, losing only a small fraction of electrons and positrons by pair annihilation. It is conceivable that life could adapt itself to such an austere mode of existence. But I have chosen to ignore this possibility. I shall assume that matter is permanent and that life can take advantage of the diversity of physical and chemical processes that matter provides. In any case, even if this assumption is wrong, it is certainly good for the next 10^{33} years, long enough for us to study the situation carefully. Long before the protons disappear, life may well be extinguished by other physical or cosmological contingencies.

Even if survival is not forbidden by the known laws of nature, the question remains open whether survival is possible for real life in the real world. The first person who thought seriously about this question was Jamal Islam, who wrote a paper about it in 1977 and later wrote an excellent book discussing it in greater detail (Islam 1977, 1983). I started to think about it after reading Islam's paper and gave some lectures that were published in the *Reviews of Modern Physics* in 1979 with the title "Time without End" (Dyson 1979). I was amazed that the *Reviews of Modern Physics* accepted my speculations in a journal that is usually considered to be respectable. I arrived at an optimistic conclusion, that the laws of nature do not make it impossible for life to survive forever. After that came a paper by Steven Frautschi (1982), also reaching an optimistic conclusion. And then, silence for fifteen years. The question of ultimate survival disappeared from the physics journals. It was no longer respectable science and was left as a theme for writers of science fiction to embroider. Writers of science fiction are allowed to violate the laws of nature if that helps to make a good story. So I stopped thinking seriously about the question.

A sudden change came in the year 1999. Two scientists at Case Western Reserve University in Cleveland, Lawrence Krauss and Glenn Starkman, sent me a paper with the title "Life, the Universe, and Nothing" (1999a). This is a serious piece of work, the first new and important contribution to the subject since 1983. It is solid science and not science fiction. And it says flatly that survival of life forever is impossible. It says that everything I claimed to prove in my *Reviews of Modern Physics* paper is wrong. I was happy when

I read the Krauss-Starkman paper. It is much more fun to be contradicted than to be ignored.

In the years since I read their paper, Krauss and Starkman and I have been engaged in vigorous arguments, writing back and forth by e-mail, trying to poke holes in each others' calculations. The battle is not over. But we have stayed friends. We have not found any holes that cannot be repaired. It begins to look as if their arguments are right, and my arguments are right too. That is a very interesting situation, when two arguments that are both right lead to opposite conclusions. It is a situation that physicists call "complementarity," when two points of view are both correct but cannot be seen simultaneously. I am beginning to think that my view of life and Krauss and Starkman's view of life are complementary. If this is true, it means that we may have together reached a deeper understanding of the question of survival than either of us had reached separately. In the rest of this chapter I will try to explain the arguments of both sides, and to show how they can lead to opposite conclusions and still both be right. Krauss and Starkman published a nontechnical version of their paper in the November 1999 issue of *Scientific American,* with the title "The Fate of Life in the Universe" (1999b).

What Do We Mean by Life?

The arguments about survival only make sense if we take a very broad view of what life means. If we take a narrow view, supposing that life has to be made of flesh and blood, with cells full of chemicals dissolved in water, then life certainly cannot survive forever. Life based on flesh and blood, the kind of life we know at first hand, can only exist at temperatures in the range above zero Celsius where water is liquid. It requires a constant input of free energy to keep it going. In a cold expanding universe, the available store of free energy in any region is finite, and life must ultimately run out of free energy if it keeps its temperature fixed. The secret of survival is to cool down your temperature as the universe cools. If you can stay alive while your temperature goes down, you can use energy more and more frugally. If you are frugal enough, you can keep going forever on a finite store of free energy. But this requires that life and consciousness can transfer themselves from flesh and blood to some other medium.

One of my favorite books is *Great Mambo Chicken and the Transhuman Condition,* by Ed Regis (1990). The book is a collection of stories about weird ideas and weird people. The transhuman condition is an idea suggested by Hans Moravec (1988), a well-known mathematician and computer expert. The transhuman condition is the way you live when your memories and mental processes are downloaded from your brain into a computer. The wiring system of the computer is a substitute for the nerve fibers in your brain. You can then use the computer as a backup, to keep your personality going in case your brain gets smashed in a car accident, or in case it develops Alzheimer's. After your old brain is gone, you might decide to upload yourself into a new brain, or you might decide to cut your losses and live happily as a transhuman in the computer. "When you die," said Moravec, "this program is installed in a mechanical body that then smoothly and seamlessly takes over your life and responsibilities." The transhumans won't have to worry about keeping warm. They can adjust their temperature to fit their surroundings. If the computer is made of silicon, the transhuman condition is silicon-based life. Silicon-based life is a possible form for life in a cold universe, whether or not it began with creatures like us made of flesh and blood.

Another possible form of life is the Black Cloud described by Fred Hoyle in his famous science-fiction novel of that title (1957). The Black Cloud lives in the vacuum of space and is composed of dust grains instead of cells. It derives its free energy from gravitation or starlight and acquires chemical nutrients from the naturally occurring interstellar dust. It is held together by electric and magnetic interactions between neighboring grains. Instead of having a nervous system or a wiring system, it has a network of long-range electromagnetic signals that transmit information and coordinate its activities. Like silicon-based life and unlike water-based life, the Black Cloud can adapt to arbitrarily low temperatures. Its demand for free energy will diminish as the temperature goes down.

Silicon-based life and dust-based life both require that protons should be stable. If protons are unstable, neither silicon nor dust can exist forever. If protons are unstable, there can be no permanent solid structure resembling a silicon-based computer. But a structure resembling a Black Cloud might still exist, with free electrons and positrons replacing dust grains. The processes of life might be embodied in the organized motions of electrons and positrons, just as Hoyle imagined them embodied in the organized motions of dust grains.

Silicon-based life and dust-based life are fiction and not fact. I am not saying that silicon-based life or dust-based life really exists or could exist. I am using them only as examples to illustrate an abstract argument. The examples are taken from science fiction, but the abstract argument is rigorous science. I give you the examples to make the abstract argument clearer. Without the examples, the abstract concepts are difficult to explain. But the abstract concepts are generally valid, whether or not the examples are real. The concepts are digital life and analog life. A transhuman living in a silicon computer is an example of digital life. A black cloud living in interstellar space is an example of analog life. The concepts are based on a broad definition of life that Krauss and Starkman and I have agreed to accept. Life is defined as a material system that can acquire, store, process, and use information to organize its activities. In this broad view, the essence of life is information, but information is not synonymous with life. To be alive, a system must not only hold information but process and use it. It is the active use of information, and not the passive storage, that constitutes life.

We are all familiar with the fact that there are two ways of processing information, analog and digital. An old-fashioned long-playing record gives us music in analog form, a compact disk gives us music in digital form. A slide rule does multiplication and division in analog form, an electronic calculator or computer does them in digital form. So we define analog life as life that processes information in analog form, digital life as life that processes information in digital form. To visualize digital life, think of a transhuman inhabiting a computer. To visualize analog life, think of a Black Cloud. The next question that arises is whether we humans are analog or digital. We don't yet know the answer to this question. The information in a human is mostly to be found in two places, in our genes and in our brains. The information in our genes is certainly digital, coded in the four-letter alphabet of DNA. The information in our brains is still a great mystery. Nobody yet knows how the human memory works. It seems likely that memories are recorded in variations of the strengths of connections between the billions of nerve cells in the brain, but we do not know how the strengths of connections are varied. It could well turn out that the processing of information in our brains is partly digital and partly analog. If we are partly analog, the downloading of a human consciousness into a digital computer may involve some loss of our finer feelings and qualities. That would not be surprising. I certainly have no desire to try the experiment myself.

There is also a third possibility, that the processing of information in our brains is done with quantum processes, so that the brain is a quantum computer. We know that quantum computers are possible in principle, and that they are in principle more powerful than digital computers. But we don't know how to build a quantum computer, and we have no evidence that anything resembling a quantum computer exists in our brains. Since we know so little about quantum computing, Krauss and Starkman and I have not considered it in our arguments. We discuss the possibilities for life based either on digital or on analog processing, the two kinds of information processing that we more or less understand.

Before I describe the details of the argument between Krauss and Starkman and me, let me state the conclusion. The conclusion is that they are right, and life cannot survive forever, if life is digital, but I am right, and life may survive forever, if life is analog. This conclusion was unexpected. In the development of our human technology during the last fifty years, analog devices such as LP records and slide rules appear to be primitive and feeble, while digital devices such as CDs and personal computers are overwhelmingly more convenient and powerful. In the modern information-based economy, digital wins every time. So it was unexpected to find that under very general conditions, analog life has a better chance of surviving than digital life. More precisely, the laws of physics and information theory forbid the survival of digital life but allow the survival of analog life. Perhaps this implies that when the time comes for us to adapt ourselves to a cold universe and abandon our extravagant flesh-and-blood habits, we should upload ourselves to black clouds in space rather than download ourselves to silicon chips in a computer. If I had to make the choice, I would go for the black cloud.

The superiority of analog life is not so surprising if you are familiar with the mathematical theory of computable numbers and computable functions. Marian Pour-El and Ian Richards, two mathematicians at the University of Minnesota, proved a theorem twenty-five years ago that says, in a mathematically precise way, that analog computers are more powerful than digital computers (Pour-El and Richards 1981). They give examples of numbers that are proved to be noncomputable with digital computers but are computable with a simple kind of analog computer. The essential difference between analog and digital computers is that an analog computer deals directly with continuous variables while a digital computer deals only with discrete variables. Our modern digital computers deal only with zeroes and ones. An idealized version of an analog computer represents a continuous variable by a classical field propagating though space and time. The classical electromagnetic field obeying the Maxwell equations would do the job. Pour-El and Richards show that the field can be focused on a point in such a way that the strength of the field at that point is not computable by any digital computer, but it can be measured by a simple analog device. The idealized situations that they consider have nothing to do with biological information. The Pour-El–Richards theorem does not prove that analog life will survive better in a cold universe. It only makes this conclusion less surprising.

Digression into Cosmology

The possibility of survival depends on cosmology as well as on the nature of life. Modern cosmologists have given us a wide choice of cosmological models that might describe the universe we live in. Astronomers are working hard to compare these models with the evidence obtained from measurements of the cosmic microwave background radiation, from studies of gravitational lenses, from the statistical distribution of clusters of galaxies, and from the statistics of catastrophic events such as supernovas and gamma-ray bursts. I won't try to cover all the possibilities. I will mention only two model universes that have the virtue of being simple. The two models have radically different consequences for the future of life.

It is convenient to use the names linear and accelerating to describe the two models. Which of them we are living in depends on the nature of the dark invisible energy that permeates the universe and outweighs the visible and invisible matter by a factor of two. The linear universe is the model that I used as the basis of my analysis of the survival problem in 1979. It has the universe expanding linearly with time forever. The relative velocity of any pair of distant galaxies remains constant. This means that we cannot exchange significant quantities of matter or energy with distant galaxies but can continue forever to exchange significant quantities of information. Every life-form must survive using the finite resources of matter and free energy contained in its local neighborhood, but has access to information from all over the universe. Until a few years ago, the linear universe was generally believed by astronomers to be the correct model. It was consistent with the older observations, which were not very accurate and did not extend far back into the past.

The accelerating model of the universe emerged a few years ago as a serious possibility, as a result of new observations of the microwave background radiation and of distant supernovas. The new observations are wonderfully accurate and go back almost all the way to the beginning of the universe. The accelerating model has the universe expanding exponentially rather than linearly. The expansion is driven by an exotic form of dark energy exerting a repulsive rather than an attractive force on the visible and invisible matter. Every distant galaxy moves away from us with increasing velocity, until after a finite time it disappears over the horizon of the universe and is no longer visible. This is a dismal situation for life. Every living community will ultimately be isolated within a finite volume of space, with no possibility of communicating with the rest of the universe. In the sky we will see the fading images of stars belonging to our own local group of galaxies, and beyond that, absolute blackness. No trace of a signal coming from more distant worlds. Krauss and Starkman and I agree that in these circumstances life cannot survive forever.

Almost all professional astronomers are now convinced by the evidence from microwave background radiation and supernovas that the universe is accelerating. By this they mean that the distant galaxies are accelerating away from us at the present epoch of cosmic evolution, when the age of the universe is about fourteen billion years. But this still leaves open two possibilities for the expansion of the universe in the future. Either the acceleration is built into the geometry of the universe, which means that it stays constant forever and the simple accelerating model is correct for the future. Or alternatively, the acceleration may be driven by some other mysterious force that causes acceleration but does not remain constant as the universe evolves. If the acceleration is driven by an unknown force, then it may continue with diminishing strength for a few tens of billions of years and afterward fade away. The universe of the remote future will then resemble the linear model, with galaxies receding at constant velocity. Although the universe is known to be accelerating at the present time, the linear model may in the end turn out to be correct.

I like to keep an open mind about the accelerating model of the universe because there is a long history of astronomers believing confidently in some particular cosmological model and then changing their minds later. When I was a student fifty years ago, everyone believed in a closed model that had the universe collapsing into a big crunch after a finite time. Then, twenty years ago, on the basis of slender evidence, everyone believed in a decelerating model that had distant galaxies moving away from us more and more slowly as time went on. Then, ten years ago, on the basis of stronger evidence, everyone believed in the linear model. And now, on the basis of evidence collected during the last five years, everyone believes in the accelerating model. Astronomers have a tendency to follow the latest fashion. Although the latest fashion today is supported by strong new evidence, I still look at it with a skeptical eye.

The Argument for Survival

I now go back to the argument in my 1979 *Reviews of Modern Physics* paper, which claimed to show that life can survive forever using a finite amount of material and a finite amount of free energy. The argument is simple and depends only on the thermodynamic relation between free energy, temperature, and information. This relation says that energy consumed equals information processed times temperature. The processing of information always goes together with an increase of entropy, so that the amount of information processed is equal to the amount of entropy generated by the living system. The living system may be a single creature or a whole ecology of creatures.

If the life is conscious, it will experience the passage of time at a faster or slower rate depending on its rate of processing of information. The subjective time experienced by the consciousness is measured by the processed information rather than by the physical time. For life to survive forever with a finite reserve of free energy, it is only necessary that the total information processed should be infinite while the total energy consumed should be finite, when we add things up over the whole of our future history. This will be possible if the life-form reduces its rate of processing information to keep in step with its falling temperature. But to reduce the rate of processing information means to reduce the quality of life. It means that we move and think more and more sluggishly as the temperature falls.

How big a reduction in the quality of life is tolerable? This question is subject to debate. Many of us would say that a big reduction is too high a price to pay for survival. But luckily we have an escape from the dilemma. The escape is hibernation. Like bears and squirrels, any life-form can adapt to decreasing temperature by interposing periods of activity with periods of sleep. For example, a society might decide to enjoy active periods with constant activity and constant duration, interposed with sleeping periods whose duration increases with time. The temperature continues to fall slowly by radiation of energy into space during the sleeping periods, but the life-form is awake only during the active periods and does not experience any diminution of the quality of life. During the active periods, the society is living beyond its means, accumulating entropy at an unsustainable rate. If the rise in temperature caused by its activity were not halted, it would choke itself to death, but it escapes into hibernation and cools off before the next bout of activity begins. The total consumption of free energy is finite, and the enjoyment of an infinite number of active periods is possible, provided that the sleeping periods are long enough.

There are two physical limitations on the possible rate of cooling of a living system. First, the temperature cannot be less than the ambient temperature of the universe. In the linear model of the universe, the ambient temperature decreases inversely with time, so the temperature of life cannot decrease more rapidly than inversely with time. This limitation is a mild one, but the second limitation is more stringent. The temperature of a living community can only be lowered by radiating energy into space, and the rate of radiation of energy is limited by the Stefan-Boltzmann law, which says that the rate is proportional to the fourth power of temperature. As the temperature falls, the efficiency of radiation falls much faster. This means that the sleeping periods must occupy an increasing fraction of the time as the temperature falls. The community can cool itself sufficiently if the duration of the nth sleeping period increases with the fourth power of n while the duration of the nth active period remains constant. The community may miss hearing the exciting news that arrives from neighboring communities during the long night, but can store the messages and respond to them during the following day.

It is easy to make a rough estimate of the quantity of free energy required to keep a hibernating society of a given size alive forever. The free energy required to survive forever is roughly twice the present consumption rate multiplied by the time until the first hibernation. For example, our own society consumes energy at a rate of about 10^{14} watts, and it might consider going into hibernation when the sun dies about five billion years from now. The free energy required for our permanent survival is then roughly 10^{24} watt-years, which is the amount of energy radiated by the sun in three days. If we can learn to store even a tiny fraction of the sun's energy in a longer-lasting form, we shall be in a good position to deal with the energy crisis that we shall face when the sun dies, and with other more severe energy crises that may come later.

The Counterarguments

I have sketched for you the argument for survival. Now I will sketch the counterarguments raised by Krauss and Starkman. There are three main counterarguments, which I will call the quantized-energy argument, the alarm-clock argument, and the eternal repetition argument. The quantized-energy argument says that any material system, whether living or dead, must obey the laws of quantum mechanics. If the system is finite, it will have only a finite set of accessible quantum states. A subset of these states will be ground states with precisely equal energy, and all other states will have higher energies, separated from the ground states by a finite energy gap. If the system could live forever, it would ultimately become so cold that the temperature would be much smaller than the gap, and the states above the gap would become inaccessible. From that time on, the system could no longer emit or absorb energy. It could store a certain amount of information in its permanently frozen ground states, but it could not process the information and it could not reduce its entropy by radiating away energy. It would be, according to our definition, dead.

The alarm-clock argument is concerned with the machinery that allows the system to become active after a period of hibernation. There must be some kind of clock that keeps track of time during hibernation and gives the wake-up signal when the moment for resuming activity has arrived. The alarm clock must satisfy stringent requirements. It must not consume any significant quantity of free energy during hibernation. It must be reset at the end of each active period. The wake-up signaling and the resetting must be done an infinite number of times. And the energy cost of wake up and resetting must diminish fast enough so that the total cost of the infinite series of operations is finite.

The quantized-energy argument applies to the alarm clock and shows that the clock cannot work if it is included as a fixed part of a finite system. The clock must be a detached mechanism, separated from the system that it is controlling by distances that increase with time. For example, my first proposal for an alarm clock consisted of two small masses orbiting around a big mass. The two small masses were unequal, and the larger one was further than the smaller one from the big mass. The clock was set by starting the two small masses in circular orbits in the same plane about the big mass. The orbital motions generated gravitational waves that caused the orbits to slowly contract. The orbit of the larger mass contracts faster. After a long but predictable time, the orbital radii will be equal and the two small masses will collide. The collision gives the wake-up signal. And at the end of the active period the clock is reset with the distances between the three masses increased. Since the time required for gravitational waves to shrink an orbit is proportional to the fourth power of the distance, the times can easily be made long by adjusting the distances. And if the distances increase rapidly enough, the total gravitational energy that must be supplied to reset the clock an infinite number of times will be finite.

Krauss and Starkman criticized this alarm-clock proposal on several grounds. First, they said, the collision of the two small masses will dissipate a finite amount of energy. Second, the triggering of the wake-up signal after the collision will require a finite amount of energy. Third, the separation of the two small masses to reset the clock will require a finite amount of energy. All these amounts of energy remain roughly constant and do not diminish to zero as the cycle of triggering and resetting is repeated. Therefore the clock requires an infinite amount of free energy to keep working forever. The alarm clock does not help the system use energy more economically but only makes things worse.

The third argument of Krauss and Starkman is a simple one. They said that since the living system ultimately has only a finite number of accessible quantum states, it must from time to time come back to a state that it has occupied before. Its conscious memories will then be exactly the same as they were before, so that it can have no awareness that it is repeating an earlier experience. From that time on, its life will be an eternal repetition, going through the same sequence of states over and over again. No new memories and no new experiences will ever happen. This state of affairs is not what we have in mind when we speak of eternal life. Eternal repetition is not eternal life, it is eternal boredom.

Krauss and Starkman thought they had dealt a fatal blow to my survival strategy with their three arguments, the quantized-energy argument and the alarm-clock argument and the eternal repetition argument. But I am still on my feet, and here is my rebuttal. The quantized-energy argument is only valid for a system that stores information in pieces of solid matter with fixed sizes, or in devices confined within a volume of fixed size, as time goes on. In particular, it is valid for any system that processes information digitally, using discrete states as carriers of information. In a digital system, the energy differences between discrete states remain fixed as the temperature goes to zero, and the system ceases to operate when the temperature is much smaller than the energy differences. But this argument does not apply to a system based on analog rather than digital devices. For example, consider a living system like Hoyle's Black Cloud, composed of dust grains interacting by means of electric and magnetic forces. After the universe has cooled down, each dust grain will be in its ground state, so that the internal temperature of each grain is zero. But the effective temperature of the system is the kinetic temperature of random motions of the grains. The information processed by the system resides in the nonrandom motions of the grains, and the entropy of the system resides in the random motions. The entropy increases as the information is processed. But in an analog system of this kind, there is no ground state and no energy gap. The essential difference between a Black Cloud and a silicon computer is the fact that the cloud expands as its temperature cools. Since electric and gravitational energies vary inversely with distance, the linear size of the cloud varies inversely with its temperature.

To see why the quantized-energy argument fails, consider the number of quantum states available to a grain in the cloud. The number of available quantum states increases with the volume of the cloud. The quantized-energy argument is not valid for an analog system, because the energy gaps between quantum states get smaller as time goes on. At late times quantum mechanics becomes irrelevant and the behavior of the cloud becomes essentially classical. The number of quantum states is so large that classical mechanics becomes almost exact. When the grains in the cloud are behaving classically, the quantized-energy argument does not apply to them.

I rebut the alarm-clock argument in a similar fashion. I imagine the alarm clock as before, a set of three masses emitting gravitational waves as they orbit, but each mass is now a little black cloud instead of a solid body. During the hibernation periods when the clock is running, the three black clouds are asleep, like the rest of the system, radiating away their internal entropy and processing no information. When the two small masses come together, they do not physically collide but interpenetrate each other. The interpenetration disturbs them enough to wake them up and cause the alarm signal to be transmitted, but does not disrupt them. Since the relative velocities of the grains in each of the masses decrease with time, the energies involved in the interpenetration and signaling and resetting of the clock also decrease with time. Because the mechanism of the clock is expanding at the same rate as the rest of the system, the energy required to operate the clock an infinite number of times remains finite.

Life based on analog processes also escapes the eternal repetition argument. Since the size of the cloud increases with time, the number of accessible quantum states also increases with time. The capacity of memory and of consciousness are steadily increasing, and there is no danger that the entire system can ever return precisely to an earlier state. So I conclude that in the linear model of the universe we can build analog systems that escape all three of Krauss and Starkman's arguments.

Conclusions

Finally, let us examine the situation that life faces in the accelerating model of the universe. In this case, I agree with Krauss and Starkman that survival forever is impossible. They pointed out two extremely unpleasant features of the accelerating universe. First, there is a finite distance at which the repulsive force driving the acceleration overwhelms the gravitational attraction of a galaxy or a cluster of galaxies. That distance is a point of no return. Anything further away than that will be driven away to the horizon and vanish. This point of no return defeats my strategy of allowing the living system to increase its size indefinitely. Any permanent system must be confined within the no-return distance, and then the quantized-energy argument will apply to it. The second unpleasant feature of the accelerating universe is that its temperature does not decrease to zero but tends to a finite limit at late times. It is permeated by cosmic background radiation at a fixed temperature, and it is impossible for any living system to cool itself to a temperature below that. This means that the free energy required to process any amount of information is ultimately proportional to the quantity of information. If the reserve of free energy is finite, the total quantity of information is also finite. This is a dismal situation, but if the universe is permanently accelerating there is no escape.

Here is a quick summary of our conclusions. We have considered two simple models of the universe, linear and accelerating. In the permanently accelerating universe, we all agree that survival is impossible. In the linear universe, I say that survival is possible for analog life, while Krauss and Starkman say that survival is impossible for digital life. In other words, survival is possible in the domain of classical mechanics where analog machines can work, but impossible in the domain of quantum mechanics where everything is digital. Fortunately, classical mechanics becomes dominant in the linear cosmology as the universe expands and cools. But our arguments about the possibility of analog life continue, and Krauss and Starkman have not yet conceded. I am still expecting them to come back with new arguments that I will then do my best to refute.

6
Looking for Life

Where Are They?

Long ago, when the search for life in the universe was just beginning, the physicist Enrico Fermi asked the question: Where are they? He meant, if the universe is really populated with alien creatures, why do we not see them? The entire universe outside Earth, so far as we can see, looks dead. We see dead planets, dead moons, dead asteroids, and dead stars. We see life only on our own planet, where a marvelously diverse and abundant collection of species live together in a small space. The contrast between the living Earth and the dead universe is to me a call to action. How much more beautiful the universe would be if these dead planets and moons were brought to life! A dead planet may sometimes be beautiful, but the same planet teeming with bees and birds and butterflies would be far more beautiful. During the next hundred years, when our mastery of biotechnology allows us to create new species to enrich our own planet, we will also learn how to breed new species of plants and animals and microbes adapted to survive in strange environments that were never seen on Earth. We will design warm-blooded potatoes to grow wild on Mars, or butterflies to fly in space using the pressure of sunlight on their wings to navigate.

When we plant potatoes on Mars or butterflies among the rings of Saturn, we will not be dealing with single species but with whole ecologies. Biotechnology works with ecosystems as well as with genomes. Wherever life is planted, it must come with a large assortment of microbes and plants and animals to sustain itself and create its own environment. Once it is planted, it will continue to evolve by the normal processes of natural variation and selection. Humans may be the midwives to bring life to birth in distant places, but once life is on its way it will grow and evolve and spread without our help. Some of us may choose to go along with it and share the adventure of colonizing the universe. But to me it is not the human settlement of space that is important. It is life as a whole that I wish to see expanding.

The origin of life is the deepest mystery in the whole of science. Many books and learned papers have been written about it, but it remains a mystery. There is an enormous gap between the simplest living cell and the most complicated naturally occurring mixture of nonliving chemicals. We have no idea when and how and where this gap was crossed. We only know that it was crossed somehow, either on Earth or on Mars or in some other place from which the ancestors of life on Earth might have come. If we can understand how life began, we shall also have gained a deeper understanding of what it means to be alive. There will probably be no single road to understanding. We must explore on at least three of the frontiers of science: first, the chemical frontier, exploring the chains of chemical reactions that might bridge the gap between living and nonliving; second, the environmental frontier, exploring the history of the various environments on the earth and elsewhere in which life might have germinated; third, the organizational frontier, exploring the deep structure of living cells and the organization of the intricate processes that give them integrity and durability.

Until we understand how life began, we have no way to calculate how likely or unlikely it may be for indigenous life to exist elsewhere in the universe. Carl Sagan and some other famous astronomers believed strongly that life is abundant on millions of planets all over our galaxy. Biologists generally tend to believe that life is a rare accident, unlikely to be abundant and possibly existing nowhere else except on Earth. No matter whether life turns out to be abundant or rare, we shall continue searching. If we ever discover an alien life-form that originated independently, that will be an enormous help to us in understanding how life began. If we can ever catch a glimpse of an alien creature swimming in an alien ocean, we will know that life has the potential to originate and evolve wherever the conditions are right, all over the universe.

Pit Lamping at Europa

In this chapter, we come down from the lofty realms of speculation and talk about tools. The tools for exploring the universe are telescopes and spacecraft. These are also our tools when we search for life in our little corner of the universe. Most likely, planet Earth is the only place in our solar system where life exists. But it would add immensely to our understanding of life if we could find it somewhere else. I am proposing a new way to look for life in unlikely places. Searching for life can be remarkably cheap and simple if it is done the right way. The chance of finding life may be small, but it is worth a try if it can be done without breaking the bank.

I leave it to my readers to decide whether this proposal is a piece of bad science fiction or a serious contribution to the planning of future space missions. I described the proposal in 1999 to an audience of space engineers at the Jet Propulsion Laboratory in California. The JPL engineers were planning a mission that would place a spacecraft in orbit around Europa, the satellite of Jupiter that is known to possess a deep ocean covered by a crust of floating ice with large and conspicuous cracks. The main purpose of my proposal was to suggest a new way for a spacecraft orbiting Europa to look for evidence of life. The Europa orbiter mission was not funded, and the JPL engineers have turned their attention to other projects. My proposal remains on the table for future missions to the outer regions of the solar system.

My son George, who lived for many years in Canada, informs me that in rural Canada, nighttime hunting with a carefully shielded flashlight is called *pit lamping.* A pit lamp is a lamp that miners carry strapped to their foreheads when they work underground. Pit lamping is illegal because it is too effective at turning up game. The reason why it is effective is that the eyes of animals staring into the light act as efficient reflectors. The eye reflects a fraction of the light that is focused onto the retina, and the reflected light is again focused into a narrow beam pointing back to the flashlight. If you are standing behind the flashlight, you can see the eyes as bright red points even when the rest of the animal is invisible. Pit lamping is too easy and too efficient. It results in extermination of the game, spoiling the fun for other hunters. The same strategy is used legally by sheep farmers in New Zealand to exterminate rabbits who compete with sheep for grazing land. The farmers drive over the land at night in heavily armed Land Rovers with headlights on, shooting at anything that stares into the headlights and does not look like a sheep.

I am suggesting that pit lamping would be a good strategy for us to use when we are searching for life in the solar system. There are two kinds of life that we may search for, intelligent and unintelligent. At the end of the chapter, I will have something to say about searching for intelligent life. Intelligent life means advanced extraterrestrial civilizations that might be detectable over interstellar distances if they transmit high-powered communications that we can intercept. Advanced civilizations certainly do not exist in our local solar system. Within the solar system, we must search for unintelligent life, the sort of life that existed on this planet before humans came along. Pit lamping is a good way to find unintelligent life.

Consider for example the problem of looking for life on Europa. If life exists on Europa, it is most likely that it exists in the ocean under the ice. The ice is many kilometers thick. To look for life in the ocean will be very difficult and expensive. It would be much easier and cheaper to look for life on the surface. Life on the surface is less likely to be there, but easier and cheaper to detect. When we are choosing a strategy to look for life, detectability is more important than probability. Estimates of the probability of life existing in various places are grossly unreliable, but estimates of detectability are generally reliable. Probability depends on theories of the origin of life, about which we know nothing, while detectability depends on capabilities of our instruments, about which we know a lot. So I state as a general principle to guide our efforts to search for life: go for what is detectable, whether or not we think it is probable. For all we know, all kinds of unlikely things might be out there, and we have a chance to find them if and only if they are detectable.

When we are looking for life on Europa, we must begin by considering the question: What kind of life could exist and survive on the surface of Europa, where the ambient temperature is about minus 150 Celsius? We may imagine that life originated and evolved for a long time where it is warm, in the dark waters below the ice. Then by chance some living creatures were carried upward through cracks in the ice, or evolved long shoots pushing up like kelp through the cracks, and so reached the surface, where energy from sunlight was available. In order to survive on the surface, the creatures would have to evolve little optical mirrors concentrating sunlight onto their vital parts. Some Arctic plants on planet Earth have evolved something similar, with parabolic petals reflecting sunlight onto the ovaries and seeds that grow at the center of the flower. On Europa the mirrors would have to be more focused than on the Arctic tundra of Earth, but not by a big factor. Sunlight on Europa is twenty-five times weaker than sunlight on Earth, so the mirrors on Europa would have to concentrate sunlight by a factor of twenty-five. That would be sufficient to maintain the illuminated area at the focus of the mirrors at a temperature above zero Celsius, at which water is liquid and life as we know it can function. The mirrors would not need to be optically perfect. Concentration of sunlight by a factor of twenty-five could be done with a rough approximation to parabolic shape.

Now suppose that we have a Europa orbiter looking for life on Europa. If the life happens to be on the surface, it must have some kind of reflectors to concentrate sunlight. Then all we have to do is to let the orbiter scan the sunlit face of Europa, always pointing the camera in a direction directly away from the sun. If there is an optical concentrator on the surface, any sunlight that is not absorbed at the focus will be reemitted and reflected into a beam pointing back at the sun. Like the eyes of a moose in Canada or the eyes of a rabbit in New Zealand, anything alive on the surface will be twenty-five times brighter than its surroundings, assuming that the illuminated surface at the focus of the concentrators has the same reflectivity as the surroundings. Even if the surface at the focus has lower reflectivity, so long as it is not jet black, any patch of life on the surface will stand out from its surroundings like the animals in a pit-lamping hunt. Since our purpose is not to kill and eat the creatures but only to discover them, there is no reason why pit lamping on Europa should be illegal.

If you sit in a window seat on the shady side of a commercial airliner and look down at a layer of cloud not too far below, you can sometimes see a beautiful bright halo surrounding the shadow of the airplane as the shadow moves along the clouds. The halo is produced by backscattering of light in water droplets or ice crystals in the cloud. This is only one of many ways in which nonliving materials can mimic the backscattering effect of a living creature. The fine dust on the surface of our earth's moon also produces a sharp increase in brightness when the moon is observed with the sun directly behind the earth. If we observe strong backscattering of sunlight from a patch on the surface of Europa, we should not immediately claim that we have discovered life. Other explanations of the backscattering must be carefully excluded before any claims are made. A good way to do this would be to carry a sensitive infrared camera looking in the same direction as the optical camera. If the optical backscattering is caused by reflection from optical concentrators attached to living creatures, then the warm surface at the focus of the concentrators will radiate thermal infrared radiation, and the thermal radiation will also be reflected back in a narrow beam pointed toward the sun. The infrared camera will detect a signal from a patch of living creatures viewed from the direction of the sun, and this signal will stand out from the cold surroundings even more strongly than the optical signal. It is unlikely that any nonliving backscatterer could mimic this infrared signal. Another way to distinguish living from nonliving backscatterers would be to analyze the backscattered light with a spectroscope and look for spectral features that might be identified with biologically interesting molecules. Laser illuminators of various wavelengths could also be used.

The same strategy of orbital pit lamping could be used to search for evidence of life on any celestial body that has a cold surface with a transparent atmosphere or no atmosphere at all, and is situated far from the sun but not too far for us to send an orbiter. After we are done with Europa, we could try the same technique on other moons of Jupiter, on any of the moons of Saturn except Titan, on the moons of Uranus and Neptune, on the Trojan asteroids, or on asteroids in the main belt between Mars and Jupiter.

But this would not be a promising way to go, if the only purpose of the orbiters were the search for life. After the first two or three orbiters had failed to find evidence of life, it would be hard to get funding for more of the same. It should be a fundamental principle in the planning of space operations that every mission searching for evidence of life should have other important scientific objectives, so that the mission is worth flying whether or not it finds evidence of life. We should never repeat the mistake that was made with the Viking missions to Mars in 1976, whose advertised purpose was to give a definitive answer to the question whether life exists on Mars. After the Viking missions failed to find evidence of life, the further exploring of Mars was set back for twenty years. We should be careful to see that every orbiter dispatched to the outer solar system takes full advantage of the opportunity to explore a new world, whether or not the new world turns out to be inhabited. Fortunately, there is another promising and cost-effective way to search for life beyond Europa. The second way is to use pit lamping, not from orbit close to the target but from back here on Earth. The second kind of pit lamping, which I call home-based pit lamping, can be done from telescopes like Hubble orbiting around the earth or, even better, from big telescopes on the ground.

Home-Based Pit Lamping

When we do home-based pit lamping, we make a big jump in the distance and the number of objects we might hope to search. Think of the Kuiper Belt, which begins about ten times further from the sun than Europa and is supposed to contain billions of cometary objects of size ranging from a kilometer upward. Suppose that life has somehow succeeded in establishing itself on the surface of one of these objects. This may seem unlikely, but nobody can prove it to be impossible. Let us imagine that the surface of one of the objects at the inner edge of the Kuiper Belt is covered with living creatures. These creatures are living in sunlight that is a hundred times weaker than sunlight on Europa. So they must have evolved optical concentrators that are a hundred times stronger. They must concentrate sunlight by a factor of twenty-five hundred instead of twenty-five. This still does not require high-precision optics. The angular resolution of the concentrators must be of the order of one degree instead of ten degrees. This is still about fifty times less precise than the optics of the human eye. A roughly parabolic reflecting surface would be good enough to do the job. When sunlight falls upon the concentrator, the light that is not absorbed is reflected into a beam one degree wide that is pointed back toward the sun. Now it happens that the earth's orbit around the sun is also about one degree wide when seen from the inner edge of the Kuiper Belt. The earth is close enough to the sun so that the reflectors will increase the brightness of the object as seen from the earth by a factor on the order of twenty-five hundred. As the earth moves around the sun, when we look outward at an object in the Kuiper Belt covered by living creatures, the brightness of the object will vary noticeably as the earth moves nearer or farther from the axis of the reflected beam. It will be brightest when the earth is directly between the sun and the object.

Home-based pit lamping is not a substitute for orbital pit lamping. The two kinds of pit lamping complement each other without much overlap. Both are needed if we are to extend our search for life as widely as possible. Orbital pit lamping can search for small and sparse patches of life on a few objects such as Europa. Home-based pit lamping can search for big patches of life covering the surfaces of more distant objects. Orbital pit lamping on Europa is like searching for lichens in rocks in Antarctica with a handheld magnifying glass. Home-based pit lamping in the Kuiper Belt is like searching for rain forests on Earth with a camera on Mars.

To discuss the possibilities of home-based pit lamping in a quantitative way, I need to estimate the brightness of the reflectors. Since we are talking about astronomical observations, it is convenient to use the astronomical unit of brightness, which is the magnitude. The magnitude is a logarithmic measure of brightness. Bigger magnitude means smaller intensity of light. For those of you who are not amateur astronomers, here are the magnitudes of the planets in the outer part of the solar system: Jupiter minus two, Saturn plus one, Uranus six, Neptune eight, Pluto fifteen. Adding one to the magnitude means making the light fainter by a factor of two and a half. Now suppose that Pluto itself were covered with living creatures with optical concentrators. Then its light would be brightened by a factor of twenty-five hundred, which means eight magnitudes. Pluto would be magnitude seven, almost as bright as Uranus and brighter than any of the asteroids except for Vesta. Pluto would have been discovered two hundred years ago when the first asteroids were found. So we can be sure that Pluto is not covered with exotic sunflowers gazing at the distant sun.

Pluto is a unique object, the brightest in the Kuiper Belt, with diameter three thousand kilometers and magnitude fifteen. The faintest objects we can identify in the Kuiper Belt with ground-based telescopes are about a thousand times fainter, with magnitude twenty-three and diameter about a hundred kilometers. With the Hubble telescope we can detect objects about five magnitudes fainter, to magnitude twenty-eight and diameter ten kilometers, but we don't usually get enough observing time on Hubble to determine the orbit of an individual object and identify it as belonging to the Kuiper Belt. The practical limit for identifying Kuiper Belt objects is magnitude twenty-three, until we have a space telescope with large amounts of time dedicated to this purpose.

Suppose one of the smaller objects at the inner edge of the Kuiper Belt is covered with sunflowers. Suppose that it is barely detectable, with magnitude around twenty-three. That means that without the sunflowers it would have had magnitude thirty-one, sixteen magnitudes fainter than Pluto, far too faint to be detected even by Hubble. With the sunflowers, if it has the same reflectivity as Pluto, it would be fifteen hundred times smaller than Pluto. Its diameter would be about two kilometers. The home-based pit lamping strategy enables us to detect any object at the inner edge of the Kuiper Belt that has as much as three square kilometers of area covered with sunflowers. And this is not all. Beyond the Kuiper Belt there is the Oort Cloud, containing billions of objects orbiting the sun at distances extending out further than a tenth of a light-year. In March 2004 some astronomers at the California Institute of Technology announced the discovery of Sedna, the first identified object in the Oort Cloud. Sedna is almost as large as Pluto and travels on a long elliptical orbit extending from three to thirty times Pluto's distance from the sun. Pit lamping becomes more and more effective, the farther out you go.

For nonliving objects shining by reflected sunlight, the brightness varies with the inverse fourth power of distance, two powers of distance for the sunlight going out and another two powers for the reflected light coming back. For living objects with optical concentrators, the concentration factor increases with the square of distance to compensate for the decrease in sunlight. The intensity of the reflected beam varies with the inverse square instead of the inverse fourth power of distance. Because the linear size of the reflected beam is independent of the distance of the object, the earth will remain within the beam no matter how far away the object sits. If an object is at a tenth of a light-year distance, the concentration factor will be a hundred million, and the angle of the reflected beam will be about twenty seconds of arc. This requires optical quality a little better than the human eye but much less precise than an ordinary off-the-shelf amateur telescope. At that distance, we could identify an Oort Cloud object covered with sunflowers if its diameter were as large as four hundred kilometers. This diameter is much smaller than Pluto and comparable with other Kuiper Belt objects that were discovered in the last few years.

The prospects for pit lamping look even more promising if we look at the situation not from our Earth-bound point of view but from the point of view of the life that we are searching for. A community living on the surface of a small object far from the sun has two tools available to increase its chances of survival. One tool is to grow optical concentrators to focus sunlight. The other tool is to spread out into the space around the object, to increase the area of sunlight that can be collected. So far, we have only considered the effect of optical concentrators. We next consider the effect of increasing the area by spreading out leaves and branches into space.

Imagine an ecological community containing a large variety of species ranging from bacteria to plants and animals, comparable with a rain forest on Earth, and occupying one square kilometer of area. It requires for its sustenance an input of two hundred watts per square meter of sunlight, or two hundred megawatts for one square kilometer. Now imagine this community installed on a little comet of diameter one kilometer and moved to the Kuiper Belt. It must grow optical concentrators, as we have seen, to concentrate sunlight by a factor of twenty-five hundred and keep its absorbing surfaces at a comfortable temperature. But it must also increase its absorbing area by a factor of twenty-five hundred to keep its total supply of energy constant. It must grow out into space to form a disk of diameter fifty kilometers instead of one kilometer. The gravity of an object of this size is so weak that it imposes no limit on the distance to which a life-form such as a tree can grow. The living community with diameter fifty kilometers, with its optical concentrators spread over the enlarged area, will then be visible from Earth as a speck of light of magnitude fifteen, about as bright as Pluto.

Now suppose that the same living community is moved out to

the Oort Cloud at a distance of a tenth of a light-year, a factor two hundred further away than Pluto. To maintain the same lifestyle with the same population of creatures, it must concentrate sunlight by a factor of a hundred million and also increase its area to a hundred million square kilometers. It will grow into a thin disk of diameter ten thousand kilometers, covered with big flimsy mirrors to concentrate the light. The original comet with diameter one kilometer will have a mass around one billion tons. To make an optical concentrator, using thin film reflectors one micron thick, requires about one gram per square meter or one ton per square kilometer. With an area of a hundred million square kilometers, the community needs to use only one-tenth of the mass of the comet to make the reflectors, with the other nine-tenths left over to use for its other activities and for building structures. Now comes the remarkable conclusion of this calculation. The community in the outer region of the Oort Cloud is still visible to us as a magnitude fifteen object. The inverse fourth power dependence of brightness on distance, which holds for nonliving objects, can be completely canceled for living objects. Living communities can cancel two powers of distance by their use of optical concentrators, and can cancel two more powers of distance by their increase in area. If the rate of consumption of energy by a community is fixed, the brightness of the community as seen from the earth is independent of distance. We can easily detect any community that uses a hundred megawatts of sunlight or more for its life support, out to a distance of a light-year from the sun. Beyond a light-year, the sun is outshone by Sirius. Anything living beyond a light-year from the sun would point its concentrators at Sirius rather than at the sun. At that point, home-based pit lamping no longer works, unless we are thinking of alien pit-lamping astronomers in orbit around Sirius, with Sirius as their home base.

The practical upshot of these calculations is that it would make sense to search for life on the surface of objects in the outer solar system with ground-based telescopes. All we need to do is look for objects in the Kuiper Belt or the Oort Cloud with unreasonable brightness. We can tell whether an object belongs to the Kuiper Belt or to the Oort Cloud by measuring its parallax, the annual displacement of its position on the sky caused by the orbital motion of the earth, and its proper motion, the steady displacement of its position caused by its own orbital motion. Anything with parallax and proper motion appropriate to the Kuiper Belt or the Oort Cloud, and with brightness greater than expected for an object at that distance, would be a candidate for an inhabited world. If we find such objects and they do not turn out to be inhabited, they would still be important discoveries. They would be objects of outstanding size, a new family of giant comets or small planets, in a remote part of the solar system. If such objects exist, either inhabited or uninhabited, it may be possible to find them by looking at the output of existing sky surveys such as the Sloan Digital Sky Survey. The Sloan Survey recorded images of millions of objects, among which a few anomalously bright Kuiper Belt objects may be hidden, waiting to be discovered. If any such objects are found, they can then be examined more carefully with larger telescopes.

Life in Vacuum

It is interesting to think in a more general way about the possible history and evolution of life if it happens to exist away from planets. We have seen on planet Earth that life has two outstanding qualities that distinguish living from nonliving objects. These qualities are adaptability and invasiveness. During the four billion years that life has existed on Earth, it has adapted itself to an amazing variety of ecological niches with an amazing variety of ways to make a living. It has invaded the most inhospitable and inaccessible corners of the planet, from frozen rocks in Antarctica to thermal vents in the depths of the oceans. It is reasonable to expect that life, if it exists elsewhere, will still be characterized by extreme adaptability and invasiveness. In an unstable and unpredictable universe, these qualities are needed if life is to survive, either on a planet or anywhere else.

The most important invasion in the history of life on Earth was the move from the ocean onto the land. This invasion was not easy. Life was confined to the oceans for nearly three billion years before it learned how to survive on land. We do not know precisely how or when the invasion began. It probably began with microbes, followed later by plants and then by animals. The fact that the earth has an atmosphere was helpful to the invaders in at least five ways. The atmosphere protected the invaders from lethal ultraviolet radiation from the sun. It also provided shielding from cosmic rays. For the plants, it provided carbon dioxide and molecular nitrogen for their metabolism. For the animals, it provided molecular oxygen for the more rapid metabolism demanded by brains and muscles. And finally, it carried clouds and rain, which provided lifesaving water to replace the ocean that the invaders had left behind. The invaders made full use of these resources provided by the atmosphere as they slowly adapted to the new environment.

Now I am speculating that if life had originated on an airless satellite such as Europa instead of on a planet with air, the move from ocean to vacuum would not necessarily have been more difficult than the move from ocean to air. Let us consider in turn the five resources provided by an atmosphere. The shielding from ultraviolet radiation would hardly be needed on Europa, since sunlight is weak, and if shielding were needed it could be provided by a thin layer of skin or bark. The shielding from cosmic rays would not be needed if the invaders evolved the same resistance to radiation damage that many species of microbes and insects evolved on Earth. The carbon, nitrogen, and other elements required for metabolism of plants are probably available in the dark stripes that we see on the surface of Europa, where brine from the ocean underneath seems to have emerged from cracks in the ice. Water is certainly available in unlimited quantities anywhere on the surface. Finally, the oxygen required by animals poses a more difficult problem, but there are many ways in which this problem may be sidestepped. One possible way was imagined by Konstantin Tsiolkovsky more than a hundred years ago in his book *Dreams of Earth and Sky* (1895).

Tsiolkovsky described creatures living in the vacuum of space. They were not divided into plants and animals but combined the characteristics of both. He called them animal-plants. Here one of them is explaining how his metabolism works: "You see these green appendages on our bodies, looking like beautiful emerald wings? They are full of chloroplasts like the ones that make your plants green. A few of your animals have them too. Our wings have a glassy skin that is airtight and watertight but still lets the sunlight through. The sunlight dissociates carbon dioxide that is dissolved in the blood that flows through our wings, and catalyzes a thousand other chemical reactions that supply us with all the substances we need." If you are an animal-plant, the oxygen released by photosynthesis in your wings is carried in your blood to your brain and muscles, so you do not need air to supply it. If you were living on the surface of Europa, you would not have needed to invent lungs. Tsiolkovsky imagined many other clever tricks by which life might have adapted to the perils and opportunities of the vacuum environment. Lacking air to transmit sound, his animal-plants found a better way to communicate with each other. "One part of their body carries under the transparent skin an area like a camera obscura, on which moving pictures are continually playing, following the flow of their thoughts and representing them precisely. The pictures are formed by fluids of various colors which flow through a web of fine channels under the skin." Tsiolkovsky's vacuum dwellers invented television long before we did. Tsiolkovsky himself was a lonely man isolated by his deafness, and that is perhaps why he dreamed of television as a replacement for speech.

Let us suppose that life has emerged onto the surface of some body without an atmosphere and has developed a robust ecology of creatures adapted to living in vacuum. This might have happened on Europa or anywhere else in the outer solar system. Now comes the most important advantage of vacuum life as compared with air-breathing life. It is far easier for vacuum life to spread from one world to another than it is for air-breathing life. The reason is simple. If a chunk of ice is knocked off Europa by a cometary impact, and if the chunk has vacuum life on its surface, then the life has a good chance to survive the impact and continue to flourish while the chunk wanders around the solar system. The environment after the impact is not much different from the environment before. And if the chunk happens to land on another object with not too high a relative velocity, the life has a chance to survive the second impact and establish itself on another world. This transition from one world to another is far more difficult for air-breathing life. Joe Kirschvink and colleagues (2000) have shown that it is possible that life could have spread from Mars to Earth in the interior of a chunk of rock knocked off Mars and landing on Earth. He studied the patchy magnetization in thin slices of the famous Mars rock ALH84001, which was picked up in Antarctica, and demonstrated that the magnetization would have disappeared if the interior of the rock had ever been heated above forty degrees Celsius during its exit from Mars or during its arrival on Earth. If a living bacterium or spore had been hiding inside that piece of rock, it could conceivably have survived the trip and emerged to populate the earth with its descendants. But any living passengers in rocks traveling from Mars to Earth must have been microbes or spores in a state of suspended animation. Air-breathing creatures could not have come to Earth in this way. Only if life is adapted to vacuum, a whole ecology of more advanced creatures might make the trip together from one world to another.

Europa, as it happens, is one of the more difficult places for life to escape from, because of the strong gravitational attraction of Jupiter. The escape velocity from Europa itself is less than two kilometers per second, but the escape velocity from Jupiter starting from the orbit of Europa is about six kilometers per second. If a chunk of ice escapes Europa but does not escape Jupiter, it will continue to orbit Jupiter in a Europa-crossing orbit, and it is likely to crash into Europa again before it encounters any other possible destination. To escape from Jupiter with a velocity of six kilometers per second departing from Europa, it must have the good luck to leave in the right direction so that its velocity is added to the orbital velocity of Europa. But hopping from world to world becomes rapidly easier as we move out beyond Jupiter to the Kuiper Belt and the Oort Cloud.

If surface life exists on a typical object in the Kuiper Belt with diameter of a few kilometers, the escape velocity will be only a few meters per second. Orbital motions in the Belt are slow, and relative velocities of neighboring objects are typically around one kilometer per second or less. Gentle collisions between neighboring objects will be common. Life has a good chance to survive when a gentle collision knocks a piece of its home into space, and when the piece later makes a gentle collision to land on another object. In this way, life could spread from world to world like neutrons in a divergent nuclear chain reaction. A large fraction of the objects in the Kuiper Belt might become inhabited. If this should happen, the process of Darwinian evolution would then select life-forms that are particularly well adapted to traveling. Life forms could evolve that do not require a chance collision to leave their home but spontaneously hop into space when conditions at home become crowded, taking with them enough material resources to survive independently. Such life-forms would be winners in the race to colonize new worlds.

Life-forms that do not hop, but grow far out into space around their home territory, would also have an advantage in spreading to other worlds. In addition to acquiring more sunlight, they would acquire a bigger cross section for collisions with other Kuiper Belt objects. When life has grown out into a thin disk with an area thousands of times larger than its original territory, the effect of a collision will usually be to punch a small hole in the disk, with minor damage to the life that stays behind, and with a substantial chance of transferring seeds of life to the object that punched the hole. In the Kuiper Belt there is probably a substantial quantity of matter in the form of dust and ice crystals, too small and too dispersed for our telescopes to detect. By growing out into space, life-forms could greatly increase the amount of this material that they encounter, and could collect and use it for their own nourishment. Life-forms that adopt this strategy would be like the filter feeders that we find in tide pools on Earth. Other life-forms in the Kuiper Belt might adopt a more active strategy, using eyes to locate larger objects floating by with small relative velocity, then hopping into space to intercept them. When life has colonized a substantial fraction of Kuiper Belt objects and the competition for real estate becomes intense, we may expect to see life-forms diversifying into predators and prey, carnivores and herbivores. In the wide spaces of the Kuiper Belt, evolution will drive life to take maximum advantage of the occasional collisions and catastrophes that punctuate its otherwise quiet existence. Life will be driven by Darwinian competition toward maximum adaptability and invasiveness, the same qualities that competition nurtured on our own planet.

Beyond the Kuiper Belt lies the Oort Cloud, where the distances between habitable objects are larger and the relative velocities smaller. In the Oort Cloud, relative velocities will be of the order of a hundred meters rather than a kilometer per second. Here it will be even easier for life to hop from one object to another. We know very little about the total number of these objects, but we know quite a lot about their chemical and physical constitution, since these are the objects that we see as long-period comets when they are occasionally deflected by gravitational perturbations into orbits that pass close to the sun. Then we see them boil off tails of gas and dust that we can observe and analyze. We know that they typically have diameters of a few kilometers and are largely composed of the biologically essential elements, hydrogen, carbon, nitrogen, oxygen. Since the reservoir of these objects in the Oort Cloud delivers about one object per year into the inner solar system where we can observe it as a comet, and since the system is four billion years old, the reservoir must contain at least several billion objects of kilometer size.

It is not likely that the Oort Cloud, a vast desert of space with habitable oases separated by huge distances, is actually inhabited. But it might conceivably be inhabited by life-forms with sufficiently precise mirrors to concentrate sunlight by a factor of a hundred million. The remarkable fact is that the strategy of home-based pit lamping could enable us to detect such life-forms if they exist.

Beyond the Solar System

If vacuum life exists at all, there is no reason why it should be confined to our own solar system. There is a radio astronomer called Jack Baggaley at the University of Canterbury in New Zealand who uses a radar system called AMOR (Advanced Meteor Orbit Radar) to track incoming micrometeors that leave trails of ionized plasma in the upper atmosphere (Baggaley 2000). The tracking is accurate enough so that he can determine the velocities of individual objects with an error less than ten kilometers per second. The objects that he can detect are larger than normal interplanetary dust particles and smaller than normal meteorites that fall to the ground. He finds that a substantial fraction of these objects have hyperbolic velocities relative to the sun, so that they do not belong to the solar system but are coming in from outside. He finds a second fact which is not so firmly established but still appears to be true. Of the objects that come from outside the solar system, a substantial fraction are coming from a single direction, roughly from the direction of the star Beta Pictoris, known familiarly to astronomers as Beta Pic.

Beta Pic is famous because it is a bright star, only sixty-three light-years distant from the sun, with a very large disk of dust orbiting around it. The disk around Beta Pic is estimated to contain about a thousand times as much material as our Kuiper Belt. Unfortunately, Beta Pic is observed to be moving away from the sun with a velocity of twenty kilometers per second. Most of the dust grains ejected from the Beta Pic system will continue moving with Beta Pic away from the sun. But it is still possible that a small fraction of the grains will be ejected with high enough velocity and in the right direction to strike the earth. The dust grains that Baggaley observes coming from Beta Pic are our first direct evidence that objects might actually be exchanged between one solar system and another. The dust grains could be thrown out of the Beta Pic system by gravitational encounters with Beta Pic planets that we have not yet observed. And if this happens to dust grains, it can also happen to kilometer-size objects. We must expect that among the billions of objects in our Oort Cloud there will be a small population of kilometer-size interlopers passing through on their way from Beta Pic. No such interloper has been seen among the long-period comets that we observe passing close to the sun. All the comets whose orbits have been accurately measured belong to the solar system. But the total number of observed comets is small, and the absence of observed interlopers does not prove that there are no unseen interlopers passing through the Oort Cloud.

Another radio astronomer, John Mathews at Pennsylvania State University, observing with the radar telescope at Arecibo in Puerto Rico, has also found dust grains arriving from outside the solar system, but he disagrees with Baggaley about the direction from which they are coming (Meisel, Janches, and Mathews 2002). Mathews finds a substantial fraction coming from the direction of a famous object called Geminga, a supernova remnant in the constellation Gemini. Geminga is an enigmatic object, a cloud of hot gas in interstellar space, with no resemblance to Beta Pic. I do not try to resolve the dispute between Mathews and Baggaley. It seems likely that they are both observing the same stream of dust grains. Beta Pic and Geminga are at the same celestial longitude but differ in latitude. Beta Pic is close to the latitude of New Zealand, Geminga is close to the latitude of Puerto Rico. It is easy to imagine that observational bias resulted in the different identifications of the source of the dust grains. To avoid further argument, I shall arbitrarily assume that at least some of the grains are coming from Beta Pic.

If life can exist in our Kuiper Belt, then much more life may be flourishing in the bigger and denser disk around Beta Pic. And if kilometer-size objects are leaving the Beta Pic system and arriving in our solar system, it is possible that Beta Pic life is passing through our Oort Cloud all the time. It is then an interesting question whether we could detect life on Beta Pic objects in the Oort Cloud using home-based pit lamping. We assume that the Beta Pic creatures direct their optical concentrators toward the sun as soon as they come within range. The problem is that the Beta Pic objects, unlike the Oort Cloud objects, are moving at high velocities transverse to our line of sight. Sunlight that impinges on a Beta Pic object will not be reflected straight back at the sun. The reflected light will be deflected sideways by a small angle as a result of the transverse motion. The reflected beam will miss the earth if the distance to the object is greater than a twentieth of a light-year. This leaves a very substantial volume within which detection of life on Beta Pic objects is possible.

If we are to take seriously the possibility of life traveling from one solar system to another, two more questions need to be addressed. First, is it possible for life to make the transit using ambient starlight to stay alive as it moves through the galaxy? To answer this question, we remark that here on Earth we have three stars brighter than zero magnitude in the sky, not counting the sun, namely Sirius, Canopus, and Alpha Centauri, and a couple of others, Arcturus and Vega, that are within a tenth of a magnitude of zero. If our present situation in this part of the galaxy is not exceptional, we can safely assume that on the way from Beta Pic to here we can always see at least one star in the sky that is zero magnitude or brighter. If the life in transit uses optical concentrators weighing one gram per square meter of area, then the light from a single zero-magnitude star provides two watts of available energy per hundred tons of concentrators. If the object in transit is an average comet weighing a billion tons and uses half its mass for concentrators, then the available energy is ten megawatts. This is enough energy to sustain a modest community of living creatures as they cruise across the galaxy. It would be enough to sustain a village of a few hundred human beings with a modern Western standard of living. The second question that needs to be addressed is whether, after an inhabited object arrives from Beta Pic with high velocity, it is possible for it to become a slow-moving object belonging to our own solar system. The answer to this question is again affirmative. Although the great majority of fast-moving objects will pass through our solar system without any interaction, a few will be deflected by close encounters with planets. A fraction of these few will be captured into orbits around the sun, and a smaller fraction will be deflected by further gravitational encounters into orbits indistinguishable from orbits belonging to our own system. From that point onward, the alien life would be at home in the solar system and could spread to neighboring objects as if it were native. Life adapted to vacuum has the potential to spread from its place of origin, not only from world to world within our solar system, but far and wide through the galaxy. Life adapted to an atmosphere is stuck on the planet where it started.

If we look at the universe objectively as a home for life, without the usual bias arising from the fact that we happen to be planet dwellers, we must conclude that planets compare unfavorably with other places as habitats. Planets have many disadvantages.

For any form of life adapted to living in an atmosphere, they are very difficult to escape from. For any form of life adapted to living in vacuum, they are death traps, like open wells full of water for a human child. And they have a more fundamental defect: their mass is almost entirely inaccessible to creatures living on their surface. Only a tiny fraction of the mass of a planet can be useful to its inhabitants. I like to use a figure of merit for habitats, namely the ratio *R* of the supply of available energy to the total mass. The bigger *R* is, the better the habitat. If we calculate *R* for the earth, using total incident sunlight as the available energy, the result is eighty watts per million tons. If we calculate *R* for a cometary object with optical concentrators, traveling anywhere in the galaxy where a zero magnitude star is visible, the result is, as we have seen, ten kilowatts per million tons. The cometary object, almost anywhere in the galaxy, is 120 times better than planet Earth as a home for life. The basic problem with planets is that they have too little area and too much mass. Life needs area, not only to collect incident energy but also to dispose of waste heat. In the long run, life will spread to the places where mass can be used most efficiently, far away from planets, to comet clouds or to dust clouds not too far from a friendly star. If the friendly star happens to be our sun, pit lamping gives us a chance to detect any wandering life-form that may have settled here.

Optical SETI

Besides pit lamping, there is another cheap method of searching for life in the universe. The other method, which is only good for detecting intelligent life, is optical SETI. Optical SETI means Searching for Extra-terrestrial Intelligence by looking for optical flashes in the sky. The SETI enterprise was started by Philip Morrison and Giuseppe Cocconi, who suggested in 1959 that we should try to detect alien friends and colleagues in the sky by listening to their radio signals. Frank Drake and Otto Struve lost no time in starting a radio SETI program at the Green Bank observatory in West Virginia, listening for signals that might have been transmitted by aliens in orbit around a few nearby stars. The radio SETI enterprise has continued to flourish from that time until today, with searches of ever-increasing scope and sensitivity as our data-processing technology improves. A big new radio SETI observatory is under construction at Hat Creek in the extreme north of California. But already in 1961, only two years after Cocconi and Morrison invented SETI and only one year after Charles Townes invented the laser, Townes suggested that searching for laser radiation would be another good way to find aliens (Schwartz and Townes 1961). Townes observed that laser beams would be about as efficient as radio transmitters for communication over interstellar distances. There was no strong reason for the aliens to prefer radio to laser communication. A well-balanced SETI program should cover both possibilities.

It took a long time before Townes's suggestion led to any action. The technology of lasers took a long time to catch up with the technology of radio. A few pioneers began sporadic optical SETI programs in Russia and America. Finally, Paul Horowitz at Harvard and David Wilkinson at Princeton organized an optical SETI collaboration, with a telescope at Harvard and a telescope at Princeton working together, carrying out a systematic search that is still continuing today. Sad to say, David Wilkinson died soon after launching the Princeton project, but the collaboration is alive and well. The Harvard telescope is the master and the Princeton telescope is the slave, programmed to point at the same patch of sky. Both telescopes carry optical detectors designed by Paul Horowitz. The detectors carry accurate timing, so that the time of arrival of any signal is exactly known. The idea is to look for nanosecond optical pulses, which are not produced by any known natural astronomical source. If a nanosecond pulse is detected, it must be either a genuine alien signal or a human joker playing a trick. If a nanosecond pulse is detected at Harvard and at Princeton, with the difference in the times of arrival exactly corresponding to the direction of the putative source in the sky, then it cannot be a human joker and can only be an alien.

The Harvard optical SETI program has now been running for seven years and the Harvard-Princeton collaboration for three. We have not yet seen any aliens. The results of the first five years of observations were published in the *Astrophysical Journal* (Howard et al. 2004), with a complete description of the apparatus and a thorough analysis of all the candidate events that were observed. The main source of spurious events is electrical discharge in the optical detectors, with cosmic rays passing through the detectors as a less important source. The frequency of spurious events is low enough that the probability of spurious events occurring in coincidence at the two telescopes is negligible. The *Astrophysical Journal* paper concludes: "We have found no evidence for pulsed optical beacons from extraterrestrial civilizations."

Why should we be excited about this optical SETI program, which is finally doing what Townes recommended more than forty years ago? I am excited because it is only a beginning. The program is outstandingly cheap. The Princeton part of the program is done with a one-meter telescope that sits on the campus next to the football stadium and is useless for any other kind of astronomy. Since the detectors only respond to nanosecond pulses, the lights of the town and the stadium do not disturb the observations. The maintenance of the detectors and the processing of the data are mostly done by students who are paid very little, since this is part of their education. The program is paid for out of the physics department education budget, which means that we do not need to waste time writing proposals and negotiating contracts. The people running the program at Harvard and at Princeton are only spending a small fraction of their time on it. The program is cheap in time and effort as well as in money. It is fun for the people who are doing it, and when it stops being fun it will stop. Nobody's career depends on it continuing. When it stops, other optical SETI projects with more modern detectors and better data processing will continue. So long as optical SETI is cheap, there will always be people with enough spare time and enthusiasm to do it.

This little SETI project is important for two reasons. The first reason is, it could happen that one night we will detect an alien. Nobody involved in the project seriously expects to find an alien, but still it could happen. It makes no sense to believe as a matter of faith that alien civilizations must exist, and it also makes no sense to believe as a matter of faith that alien civilizations do not exist. All we can say is that alien civilizations are rare, since we have been listening for forty-five years and have not found one. But wildly improbable and unexpected things happen all the time in astronomy, such as Jocelyn Bell's discovery of pulsars, Alexander Wolszczan's discovery of planets orbiting around a neutron star, or Saul Perlmutter's discovery of the accelerating universe. Those discoveries were no less improbable than the discovery of an alien laser signal. It happens all the time in astronomy that our understanding of the cosmos is turned upside down by some unexpected observation made by somebody working outside the mainstream.

If somebody should happen to detect an alien laser signal, many more people would observe the source in many different channels, to learn as much about it as possible. It would probably take a long time to find out much about the aliens. In the unlikely event that we were able to decipher the alien communications, not only a new era in astronomy but a new era in human history would begin. The future of science in the era of communication with alien civilizations is totally unpredictable. Alien thought processes might remain incomprehensible to us, or they might be comprehensible and make human science obsolete. Here science ends and science fiction begins.

The second reason why the optical SETI project is important is that it is typical of thousands of other small enterprises. This reason is still valid if, as we all expect, we do not detect any aliens. The future of science will be a mixture of large and small projects, with the large projects getting most of the attention and the small projects getting most of the results. Charles Townes's invention of the laser was an example of an important result produced by a small project. As we move into the future, there is a tendency for the big projects to grow bigger and fewer. This tendency is particularly clear in the field of particle physics, but it is also visible in other fields of science such as plasma physics and crystallography and astronomy and genetics, where large machines and large databases dominate the scene. The size of small projects does not change much as time goes on. The size of small projects is measured in human beings, and a small project typically consists of one professor and three or four students. Since the big projects are likely to become fewer and slower while the small projects stay small and quick, it is reasonable to expect that the relative importance of small projects will increase with time. The Harvard-Princeton optical SETI project is an extreme example of a small enterprise, and it is pointing the way for many more such enterprises in the future.

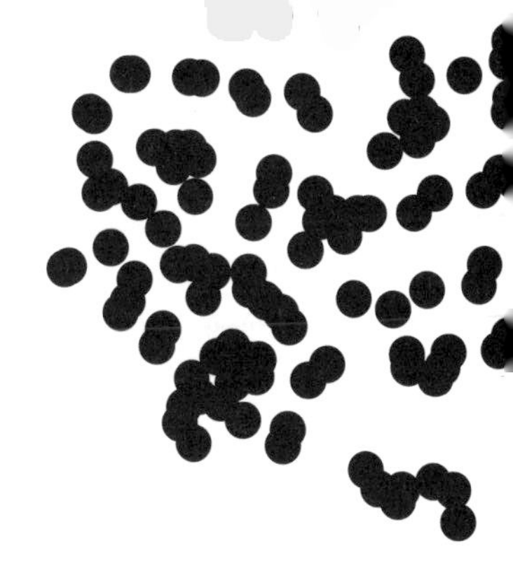

7 The Varieties of Human Experience

William James and Sir John Templeton

Religion is an essential part of the human condition, more deeply rooted and more widely shared than science. The question that I am addressing is whether our advancing knowledge of biology can lead us to a better understanding of religion. I begin with a quotation from William James's Gifford Lectures, given at Edinburgh in the years 1901–2 with the title "The Varieties of Religious Experience":

> Is the existence of so many religious types and sects and creeds regrettable? . . . I answer No emphatically. . . . No two of us have identical difficulties, nor should we be expected to work out identical

> solutions. . . . If an Emerson were forced to be a Wesley, or a Moody forced to be a Whitman, the total human consciousness of the divine would suffer. . . . Each attitude being a syllable in human nature's total message, it takes the whole of us to spell the meaning out completely. . . . We must frankly recognize the fact that we live in partial systems, and that parts are not interchangeable in the spiritual life.

My copy of *The Varieties of Religious Experience* is a little book bound in sky-blue cloth and bought in 1938 for seven shillings and sixpence; it has been my constant companion ever since (James 1937). This last chapter will be squarely based on James's way of thinking. Even the title is borrowed from James. James looked at religion from the inside. He was a professor of psychology, and he looked at religion as a psychologist looks at a patient, doing his best to see the world through the patient's eyes. He was convinced that religions are an important part of our understanding of the world, that spiritual truths exist and can be apprehended by humans, and yet he insisted that the varieties of human experience lead us to a variety of truths. There was no place for exclusiveness or for claims of infallibility in his view of religion. There was no place for dogmatic certitude. James's God was a presence that could sometimes be felt but never described. He was revealed more in people's lives than in their thoughts. James was not interested in theology, and neither was James's God.

I find it illuminating to compare William James with another great man who is embarked upon a similar quest, Sir John Templeton. Sir John recently celebrated his ninetieth birthday and is still actively engaged in supporting and encouraging the study of religion. He has a personal philosophy which he calls Humility Theology, based on the notion that we shall understand much more about God if we begin by admitting our ignorance. William James would certainly have agreed with this notion. But William James and Sir John go about the study of religion in very different ways. James studied religion by studying the individual soul. His raw material was the lives of the saints and the writings of mystics. He made no attempt to be scientific. His insights came from personal narrative, not from scientific analysis. His aim was to explore the byways of religious experience, not to reduce them to a set of scientific conclusions.

Sir John follows a different road, hoping that spiritual wisdom may be found by combining the insights of religion with the tools and methods of science. The John Templeton Foundation spends a major part of its resources in the support of research and teaching in the field of science and religion. Science and religion is a new academic discipline still in the process of defining itself. Its practitioners may be theologians, philosophers, psychologists, medical doctors, biologists, or physicists. They engage in a great variety of studies with diverse methods and purposes. But the central purpose of Sir John in supporting such studies is clear. His purpose is to rejuvenate the ancient discipline of theology by bringing into it people and ideas from the new disciplines of science. His dream is to see experts in science and religion making new discoveries in religion, as revolutionary as the discoveries that have been made during the last century in science. He uses the phrase "spiritual information" to define the goal that he is seeking.

One of the central new ideas in the physical sciences is complementarity, introduced by Niels Bohr in the 1920s as a way to describe the new world of quantum mechanics. Complementarity means the existence of two pictures of a physical process that are both valid but cannot be seen simultaneously. The best-known example of complementarity is the dual nature of light. Light sometimes behaves like a continuous wave and sometimes like a hailstorm of discrete particles. To see the wave nature of light, you do an experiment to observe its diffraction by a grating. To see the particle nature of light, you do an experiment to observe it kicking out electrons from a metal surface. The two experiments are complementary. Light is both waves and particles, but you cannot see a wave and a particle at the same time. The nature of light is richer than any of the pictures that we use to describe it.

When the idea of complementarity is applied to atomic processes governed by quantum mechanics, the idea is mathematically precise and is verified by a wealth of experiments. But Bohr liked to extend the idea to more general contexts where its use has remained controversial. Bohr introduced complementarity into biology, pointing out that a living creature can be studied either as an organic whole or as a collection of chemical molecules, but its behavior as a living organism and the behavior of its constituent molecules cannot be studied in the same experiment. In fact, the attempt to locate precisely all the molecules in a living creature would probably result in its death. He also spoke of the complementarity between justice and mercy in ethics, between thoughts and sentiments in psychology, between form and substance in literature, between frame and content in scientific theories. He spoke in an even more general way of "the mutually exclusive relationship which will always exist between the practical use of any word and attempts at its strict definition."

Following Bohr's broad use of the word, I propose that religion and science are also complementary. The formal frame of traditional theology, and the formal frame of traditional science, are both too narrow to comprehend the totality of human experience. Both frames exclude essential aspects of our existence. Theology excludes differential equations, and science excludes the idea of the sacred. But the fact that these frames are too narrow does not imply that either can be expanded to include the other. Complementarity implies exclusion. The essence of complementarity is the impossibility of observing both the scientific and the religious aspects of human nature at the same time. When we are aware of the universe through a religious experience, nothing is quantitative, and when we are aware of the universe through scientific observation and analysis, nothing is sacred. To astronomers with a religious turn of mind the heavens may proclaim the glory of God, but the glory will never be captured in their computer models of star clusters and galaxies. There is a danger that the academic discipline of science and religion may become a frame that excludes both genuine science and genuine religion. If frame A and frame B are mutually exclusive, then a frame C that tries to include both A and B is likely to end by excluding both. If science and religion are complementary, it is better that they should live apart, with mutual respect but with separate identities and separate bank accounts.

Contemporary discussions of science and religion often have a narrow focus, as if science and religion were the only sources of knowledge and wisdom. In fact, science and religion belong to a wider array of human faculties, an array that also includes art, architecture, music, drama, law, medicine, history, and literature. Several of these faculties have closer ties than science with religion. Every great religion has great art and great literature associated with it from ancient times. The connections between science and religion are by comparison recent and superficial. I find it strange that science should be singled out as the partner of religion in Sir John Templeton's vision. If we look for insights into human nature to guide the future of religion, we shall find more such insights in the novels of Dostoyevsky than in the journals of cognitive science. Literature is the great storehouse of human experience, linking together different cultures and different centuries, accessible to far more people than the technical language of science. William James was trained as a medical doctor and was familiar with the science of his time, but he paid far more attention to literature than to science in his study of religion. His book is full of marvelous quotations from writers ancient and modern, and has hardly a single reference to scientific journals.

For many years, ever since the personal computer became ubiquitous, we have heard prophets proclaiming that books will soon be obsolete, that the new generations raised on video images will no longer be interested in reading books. Nevertheless, books survive, and new books are still being written and read. Even if books become obsolete in the future, the content of books will be transferred to some other medium and literature will survive in another form. No matter how far we look into the future, humans will need a way to share stories, and the sharing of stories is the essential basis of literature. Literature enables us to share the passions of Greek and Trojan warriors in the twelfth century before Christ, and of Hebrew prophets and kings a few hundred years later. Literature will remain as the way we embalm our thoughts and feelings for transmission to our descendants. Literature survives when the civilizations that gave birth to it collapse and die. All through our history, literature and religion have been closely tied together. It is literature that gives longevity to religion. Religions that have no literature may come and go, but the Jewish Torah and the Christian Gospels and the Muslim Koran endure through the millennia. The more successful of the new religions of recent times also have their sacred books. Latter-day Saints have their Book of Mormon, Christian Scientists have their *Science and Health with a Key to the Scriptures,* and the Marxists have their holy scriptures too.

Elaine Pagels is an example of a scholar who has enlarged our view of religion by detailed study of the associated literature. She translated and elucidated the collection of ancient scrolls that were discovered hidden under the desert sand in a wine jar at Nag Hammadi in Egypt. Her book *The Gnostic Gospels* (1979) is a popular account of her work, explaining the origins of these early noncanonical Christian texts and the new light they throw on the canonical texts which later became the Christian Bible. Pagels is not a scientist. Her skills and her tools have little to do with science. She is a linguist and a historian. Her skills are intimate knowledge of the Coptic and Greek languages, and her tools are literary and historical analysis. Her work has given us a new picture of the Christian religion as it existed in early times before orthodoxies were rigidly imposed and heresies stamped out. This glimpse of a different Christianity has had great influence in broadening the scope and style of Christian thinking. It helps to free Christianity from the dogmatism of past centuries and resonates well with the new generation of students who call themselves Christian but feel more at home with heresy than with orthodoxy. The notion of complementarity can also be used to reconcile heresy with orthodoxy, to reconcile the view of Jesus seen in the Gnostic Gospel of St. Thomas with the view seen in the orthodox Gospels of the New Testament. The various Gospels give us different views, but they are views of the same Jesus.

Elaine Pagels has been a discoverer of spiritual knowledge according to Sir John Templeton's definition. I hope there will be more scholars like her, learned in the languages and histories of other cultures and other religions, who will devote their lives to discovering and interpreting other documents that were forgotten long ago or condemned as heretical. All religions have a tendency to become rigid and intolerant. Every religion has, buried in its past, heretical views that were suppressed. If we could recover some of the ancient heretical literature of other religions and make it accessible to students in the modern world, as Pagels has recovered and explained the suppressed literature of the Christian religion, we might succeed in broadening the outlook of all religions. With a broadened outlook, our diverse religions might be better able to live together in peace. Believers in each religion might come to see that all religions are complementary, giving us views of the same reality seen from different angles.

One of the finest Christian heretics was William Blake, whose poems and prophesies were not suppressed but ignored when he published them in the eighteenth and early nineteenth centuries (Blake 1939). His orthodox contemporaries considered him insane, and he narrowly escaped being put in prison for treasonable remarks against the British monarchy. Two hundred years later he is honored as an artist and as a poet and as a spokesman for the oppressed. His poem "The Everlasting Gospel" is another heretical gospel to put beside the Gospel of St. Thomas:

The Vision of Christ that thou dost see
Is my Vision's greatest enemy:
Thine has a great hook nose like thine,
Mine has a snub nose like to mine:
Thine is the friend of All Mankind,
Mine speaks in parables to the blind:
Thine loves the same world that mine hates,
Thy Heaven doors are my Hell's gates.
Both read the Bible day and night,
But thou read'st black where I read white.

In another place he wrote:

How do you know but ev'ry Bird that cuts the airy way
Is an immense world of delight, clos'd by your senses five?

William Blake, this crazy poet who invited us

To see a world in a grain of sand
And a heaven in a wild flower,
Hold Infinity in the palm of your hand
And Eternity in an hour,

gave us more spiritual information in a few lines than all the theologians and scientists of his time in their learned volumes. In the future too, if we are searching for spiritual information, we are more likely to find it among poets than among scientists.

Theology and Theofiction

In my thinking about religion, writers of science fiction play a larger role than either poets or scientists. Science fiction is generally despised both by scientists and by literary scholars as a bastard discipline combining bad science with bad writing. Much of it deserves their contempt, but some of it does not. Some of the best writing is done by a small number of writers who have adapted the style and conventions of science fiction to tell stories that have little to do with science but much to do with theology. Their stories are found on the shelves of bookstores among the classics of science fiction, but they truly belong to a different genre to which I give the name *theofiction.* Writers of theofiction present a vision that is primarily religious rather than scientific. Their characters are exploring the meaning and purpose of the universe rather than the geography of particular places. They confront age-old problems of good and evil, not paying serious attention to the astronomical vistas and technological devices that serve as stage scenery for their dramas. The writers that I shall discuss are Olaf Stapledon, Clive Lewis, Madeleine L'Engle, and Octavia Butler. I chose these four only because I am familiar with them. There are other writers of theofiction that I do not know, many of them writing in languages other than English. Perhaps the greatest works of theofiction in the literature of Europe are Dante's *Divine Comedy* and Milton's *Paradise Lost.* The limits of the genre are arbitrary, and I will not try to trace its history. I am asking whether the modern writers of theofiction may have raised important new questions that traditional theology ignored.

Olaf Stapledon is the most analytical of my four writers. When he was not writing fiction, he was a professional philosopher. At many places in his fiction, the philosopher is speaking through the mouths of his characters. The most explicitly theological of his stories is *Star Maker,* published in 1937 (Stapledon 1968). The hero is a nondescript character who sits down to rest on a hill overlooking his home and unexpectedly finds himself embarked on a tour of the universe. The first stop is a planet similar to Earth, where he finds a kindred soul to share his journey. From there he travels on to other worlds, enlarging his view of the cosmos and collecting a diverse group of fellow travelers to explore further. He travels like Dante, through realms of horror and degradation, into realms of gradually ascending philosophic calm and understanding, until he stands finally in the immediate presence of the Star Maker. He then experiences the mystical union of the cosmos with the mind of the Star Maker. But that supreme moment is tragic rather than harmonious. Like God answering Job out of the whirlwind, the Star Maker strikes him down and rejects him. The Star Maker judges his creation with love but without mercy. In the end, our entire universe, in spite of all its majesty and beauty, is a flawed experiment. The Star Maker is already busy with designs for other universes in which our flaws may be repaired.

Seven years after *Star Maker,* Stapledon wrote *Sirius,* a less ambitious but more persuasive venture into theology (Stapledon 1972). As a work of literature, *Sirius* is far superior. The story is more gripping, and the characters more finely drawn. The most memorable of the characters is Sirius, a superdog with a superhuman brain. Stapledon was writing in the 1940s, before the chemical structure of DNA and the technology of genetically modified embryos had been discovered. A modern writer writing a story about a superdog would naturally assume that the animal must be genetically modified. Stapledon did not need genetic engineering to imagine a superdog. He imagined old-fashioned growth hormones infused into the dog's brain by an old-fashioned dog breeder. The setting of the story is sheep-dog country, the hills of North Wales during the Second World War, a time and a country where dogs and humans lived and worked together with mutual respect.

The story of Sirius is a tragedy. Sirius understands both the world of dogs and the world of humans, but he can find no place for himself in either world. Searching for a place and a purpose for his life, he becomes increasingly frustrated and angry. Then, in a moment of desperation, he is overwhelmed by a religious experience. A mystical peace descends on his soul, and an awareness of God that he is unable to describe in words. Afterward he talks to his human owner and attempts to formulate a theology. The theology of a superdog is necessarily different from human theology. Sirius's God is a supreme hunter rather than a supreme judge or redeemer. Stapledon does not develop Sirius's theology in detail. Sirius's intellectual explorations are cut short, and the story ends in tragedy, because humans who do not know Sirius regard him as a dangerous monster. At the end of the story, we are left with the theological moral. God may have more qualities than we humans are capable of imagining. If we could enlarge our senses and our emotions beyond the human range, we would experience a very different God.

The second writer on my list is Clive Lewis, more commonly known as C. S. Lewis. Lewis was a professor of medieval literature and a more-or-less orthodox Christian believer. He believed in the traditional Anglican theology rather than the attenuated theology of the modern Anglican Church. For him, Satan was a real and important presence, clearly visible in the evils of twentieth-century society. Lewis became famous as the author of *The Screwtape Letters,* a collection of letters written by the senior devil Screwtape to his young nephew, instructing the beginner how to tempt humans more effectively into paths of evil. The vividly drawn character of Screwtape gives a clear impression that Lewis was more interested in Satan than in God. The character of Satan in Milton's *Paradise Lost* gives the same impression, as William Blake observed: "The reason Milton wrote in fetters when he wrote of Angels and God, and at liberty when of Devils and Hell, is because he was a true Poet and of the Devil's party without knowing it." Lewis too was a true poet and of the devil's party without knowing it.

Lewis wrote three books that are usually classified as science fiction, although they have almost nothing to do with science. These are a trilogy, *Out of the Silent Planet, Perelandra,* and *That Hideous Strength* (Lewis 1952, 1965, 1972). The three stories are staged on three planets, Mars, Venus, and Earth. In the first two stories, the alien creatures inhabiting Mars and Venus are depicted as living like Adam and Eve in the garden of Eden before the fall from grace, uncorrupted by technology and uncorrupted by evil. In Lewis's mind, technology and evil are synonymous. The main character is a philologist who is able to learn the aliens' languages and understand their cultures. His companions who travel with him from Earth are technologists who only wish to destroy and dominate. The aliens, with the help of the philologist, succeed in defending themselves and preserving the innocence of their societies against the invasion of earthly technology.

The third story, taking place on Earth, shows Satan at work on the planet that he has made his own. The battle between good and evil occurs at a university town with a strong resemblance to Oxford. The force of evil is an organization called NICE, the National Institute of Co-ordinated Experiments, which Lewis describes as "the first-fruits of that constructive fusion between the state and the laboratory on which so many thoughtful people base their hopes of a better world." The thoughtful people, for whom Lewis expresses so delicately his contempt and loathing, no doubt included many of his academic colleagues and particularly the scientists among them. The story begins with NICE successfully corrupting the leaders of the university, buying the town and the land around it, and establishing a totalitarian police force to maintain law and order. After a prolonged struggle, the forces of good, led by the philologist and other recalcitrant spirits, prevail, with substantial help from a number of mythical characters borrowed from Arthurian legend. The corrupted town and university are destroyed in an earthquake, and the story ends with virtue rewarded and lovers reunited.

When the three parts of the trilogy were published, the fairy tale *That Hideous Strength* turned out to be far more popular than the theological allegories, *Out of the Silent Planet* and *Perelandra.* Lewis then understood that his real talent was to be a writer of fairy stories, and he went on to write a succession of books, beginning with *The Lion, the Witch and the Wardrobe,* that became classics of literature for children. Meanwhile, the three books of the trilogy remain as classics in the literature of theofiction. The theology of Lewis is as conservative as the theology of Stapledon is radical. Lewis looks inward and backward to medieval England for his inspiration; Stapledon looks forward and outward to the remote future and the universe as a whole. In the preface to *That Hideous Strength,* Lewis makes a generous acknowledgment to Stapledon: "Mr. Stapledon is so rich in invention that he can well afford to lend, and I admire his invention (though not his philosophy) so much that I should feel no shame to borrow." Both Lewis and Stapledon were rich in invention, and their joint legacy is made richer by the fact that their philosophies were diametrically opposed to each other. Theofiction does not impose constraints on theology. Rather, it widens the theological imagination.

Madeleine L'Engle, the third writer on my list, has written more than thirty books of theofiction, generally classified as children's literature but widely read and admired by adults. I will not attempt to summarize them. Instead, I will concentrate on her book *Walking on Water: Reflections on Faith and Art* (1980), which gives us a direct statement of her theology. *Walking on Water* is full of memorable quotations. Here is one borrowed from Archbishop William Temple: "It is a great mistake to think that God is chiefly interested in religion." Here is one from Saint Augustine: "If you think you understand, it isn't God." Here is one from the Koran: "He deserves Paradise who makes his companions laugh." Here is one from Francis Bacon, one of the founding fathers of modern science: "If we begin with certainties, we will end in doubt, but if we begin with doubts and bear them patiently, we may end in certainty." And here is one from L'Engle herself: "I have often been asked if my Christianity affects my stories, and surely it is the other way around; my stories affect my Christianity, restore me, shake me by the scruff of the neck, and pull this straying sinner into an awed faith."

L'Engle is a Christian, but she pays more attention to scientists than to theologians. She writes:

> I had been reading too many theologians, particularly German theologians. . . . I asked questions, cosmic questions, and the German theologians answered them all. . . . I read their rigid answers, and I thought sadly, if I have to believe all this limiting of God, then I cannot be a Christian. . . . It was the scientists, with their questions, with their rapture at the glory of the created universe, who helped to convert me . . . "A Wrinkle in Time" was my rebuttal to the German theologians. . . . When I try to find contemporary twentieth-century mystics, to help me in my own search for meditation and contemplation, I turn to the cellular biologists and astrophysicists, for they are dealing with the nature of being itself, and their questions are theological ones: What is the nature of time? of creation? of life? What is human creativity? What is our share in God's work?

L'Engle's theology is Christian but not conservative. She believes in miracles, and she believes in the divinity of Jesus, but always leaves room for doubt and human fallibility. She believes also in the world soul described by William James: "There is a continuum of cosmic consciousness, against which our individuality builds but accidental fences, and into which our several minds plunge as into a mother sea or reservoir."

For L'Engle, there is a close connection between her religion and her work as a writer. She writes: "The depth and strength of the belief is reflected in the work. If the artist does not believe, then no one else will. No amount of technique will make the responder see truth in something the artist knows to be phony. My faith in a loving Creator of the galaxies . . . is stronger in my work than in my life, and often it is the work that pulls me back from the precipice of faithlessness." Although L'Engle listens to what scientists have to say and has a deep understanding of many areas of science, her theology expresses itself in literature. For her, theology is an art and not a science.

Octavia Butler, the last of my four theofiction writers, is the only one that I know personally. I once spent a day with her, talking with an enthusiastic group of children from the inner-city schools of Chicago. We were performing together as a writer-scientist team, so that I could answer questions about science and she could answer questions about everything else. She could communicate with the kids much better than I could, since she grew up poor and black in California and the kids were mostly growing up poor and black in Chicago. She knows their world from the inside. She has written two books of theofiction, with the titles *Parable of the Sower* and *Parable of the Talents* (1995, 1998). She writes about the world she grew up in, as it might be in the future if things go badly. In the books, things go very badly. The climate has changed so that Southern California gets no rain at all. The minority of rich people live in armed fortresses. The majority outside are homeless and hopeless, scavenging in the ruins of civilization. Butler's hero is Lauren Olamina, a young black woman, the daughter of a Baptist preacher. At the age of eighteen, she escapes when her home is destroyed and her family is murdered. She leads a handful of survivors on a long trek to the north. As she travels, she works out a personal religion which she calls Earthseed. She preaches Earthseed to people she meets on the way, to anybody who will listen, but the seed mostly falls on stony ground. Only a few believe and follow her. She says, "God is power, infinite, irresistible, inexorable, indifferent. And yet, God is pliable, trickster, teacher, chaos, clay. God exists to be shaped. God is Change." This is not the Baptist religion of her father. This is something new.

Parable of the Sower ends with Lauren arriving at a remote place in Northern California where she establishes a community of nine adults and four children, battle-hardened survivors of the long trek. They settle down to raise their own food and practice their own religion. By dint of hard work and self-reliance, they survive and prosper for five years. One of the community members is a medical doctor. Lauren is married to him, and they have a baby daughter. *Parable of the Talents* continues the story, with the daughter replacing her mother as narrator. After the years of chaos and banditry, a totalitarian regime is established, with a right-wing religious organization, the Church of Christian America, in charge, and shock troops of young crusaders stamping out heresy. The crusaders stamp out the Earthseed community, imprison Lauren, and take away her daughter. The daughter is adopted into a right-thinking Christian America family and has no contact with her mother for thirty years. Meanwhile, Lauren escapes from prison. She puts aside her dream of building self-sufficient Earthseed communities so long as the crusaders are raging, and instead becomes an itinerant preacher. She preaches at first secretly, and then openly as the wave of persecution subsides and her converts increase in numbers and influence.

Unknown to Lauren, her younger brother Marc also survived the destruction of their home. He too has inherited their father's talent as a preacher and is a rising star in the Church of Christian America. So brother and sister become leaders of rival religions, Marc clinging to the old orthodoxies and Lauren preaching the new wine of Earthseed. Marc keeps his talent safe in the ground while Lauren invests hers in daring ventures. An essential article of the Earthseed faith is to spread the seed of life into the universe: "We are Earthseed, and the destiny of Earthseed is to take root among the stars . . . to live and thrive on new earths . . . to explore the vastness of heaven . . . to explore the vastness of ourselves." The story ends with Lauren dying, old and rich and famous, while the first Earthseed starship is being prepared for takeoff. When it takes off, her ashes will be on board, ready to fertilize Earthseed crops on another world. Her daughter stays behind with Uncle Marc, wondering why God spurned their years of faithful service and gave that crazy old heretic his blessing.

The theology of Earthseed might be called Action Theology. It has much in common with the Liberation Theology that arose out of a similar background of poverty and oppression in Latin America. Lauren stated the essence of it in her verses: "Chaos is God's most dangerous face--amorphous, roiling, hungry. Shape chaos. Shape God. Act. . . . Only actions guided and shaped by belief and knowledge will save you. Belief initiates and guides action, or it does nothing." God exists in the shape of action, and only through action can we bring him into our lives.

I have sketched very imperfectly the four theologies that are expressed in the writings of my four authors. In Stapledon, a theology of transcendence, with a God who plays with universes as a composer plays with melodies. In Lewis, a theology of traditional Christianity, with devils and angels fighting over human souls. In L'Engle, a theology of artistic creation, with a Christian God whose nature is glimpsed in parables and stories. In Butler, a theology of action, with a God who exists to shape human action and to be shaped by human action as he evolves. What general conclusion or moral can we draw from these diverse visions of God?

We certainly cannot reach any agreed wisdom by reducing these four writers to their lowest common denominator. What they have in common is trivial. It is their profound differences that are important. Their differences teach us that theology may have wider scope and greater freedom than it has had in the past. Just as science fiction shows us that there may be more things in heaven and earth than we are capable of imagining, so theofiction tells us that there may be more different kinds of God. These four writers, and many others that I do not happen to know so well, give us glimpses of what theology may become, if theology grows to comprehend the abundant flowering of cultures and sciences and religions in the modern world. In the last thousand years, the masterpieces of Dante and Milton brought theology to millions of readers who never grappled with the writings of Aquinas or Abelard. Likewise, in the next thousand years we may expect that the next great masterpieces of theofiction will do more than the writings of professional theologians to justify the ways of God to men.

The Varieties of Neurological Impairment

The final section of this chapter is concerned with theological insights that we may derive, not from stories of imagined people, but from stories of real people who live in mental worlds different from ours. They live in different worlds because they suffer from different varieties of neurological impairment. They are truly alien intelligences living together with us on planet Earth.

We have known for hundreds of years that the universe has room in it for other intelligent inhabitants living on other planets. If our ongoing attempts to detect their existence should be successful, this will be a big triumph for science but will not be in any sense a setback for theology. Since the time of Giordano Bruno, the multiplicity of worlds has frequently been a subject for theological speculation. Isaac Newton himself remarked in one of his theological manuscripts: "And as Christ after some stay in or neare the regions of this earth ascended into heaven, so after the resurrection of the dead it may be in their power to leave this earth at pleasure and accompany him into any part of the heavens, that no region in the whole Univers may want its inhabitants" (in Manuel 1974). God may be portrayed in a million different shapes in a million inhabited worlds, without any diminution of his greatness.

Likewise, the transition from a Newtonian cosmology of absolute space and absolute time to an Einsteinian cosmology of relativistic space-time has not changed the age-old mystery of God's relation to the physical universe. I see no reason why God should be inconvenienced if it should turn out that our universe started with an unpredictable quantum fluctuation giving rise to an inflationary expansion, or if it should turn out that we live in one of a multitude of universes. My conception of God is not weakened by my not knowing whether the physical universe is open or closed, finite or infinite, simple or multiple. God for me is a mystery, and will remain a mystery after we know the answers to these questions. All that we know about him is that he works on a scale far beyond the limits of our understanding. I cannot imagine that he is greatly impressed by our juvenile efforts to read his mind. As the Hebrew psalmist said long ago, "He hath no pleasure in the strength of an horse, neither delighteth He in any man's legs" (Ps. 147). Translating the psalmist's verse into modern polysyllabic idiom, we might say, "He hath no pleasure in the teraflops of a supercomputer, neither delighteth He in any cosmologist's calculations."

We do not need to postulate alien intelligences in the sky or cosmologies with multiple universes in order to raise new questions concerning religion. Religious questions are best raised by looking at real people with real problems and real insights. Neurology comes closer than cosmology to the questions that are at the heart of theology. Neurology gives us evidence of the way human perceptions and human beliefs come into being. By studying the perceptions and beliefs of people who live in worlds different from ours, we may better understand our own. I am a physicist with no pretensions to be an expert in neurology. When I write about neurology, I write as a layman. My knowledge of neurology is largely derived, not from the technical literature, nor even from the nontechnical literature, but from television programs addressed to the general public. I have in mind four one-hour television programs that I recently watched, with the neurologist Oliver Sacks as guide. These depict in vivid fashion the four different worlds inhabited by four groups of people with different kinds of neurological impairment. They have certain features in common. Each of the four neurological impairments is congenital, each of them deprives the affected people of an important human faculty, and each of them is ameliorated by the amazing ability of the human brain to work around obstacles.

The simplest of the four syndromes is achromatopsia, the severe form of color blindness in which the color-sensing cones in the retina are missing and only the rods remain (Sacks 1998). People with achromatopsia have excellent night vision but are almost blind in direct sunlight. Many of them adapt to their disability by learning to live like nocturnal animals. Oliver Sacks showed us a community with a high incidence of achromatopsia, living on a South Pacific island. There the achromatopes specialize in night fishing, a productive occupation for which their disability turns into an advantage.

Far greater obstacles are faced by people with Usher's syndrome, who are born totally deaf and then in middle age gradually become blind. They too can adapt to their disability if they live in a supportive community. As children they become fluent in sign language. They are able to communicate with one another and to absorb an education as readily as other deaf children. Then as adults, when their sight begins to fade, they can continue to communicate by sign language, the listener touching the hands of the speaker to feel the signs. They can continue to read and write by transferring their skills from print to Braille. Within the community of the deaf-blind, they are not isolated by their double disability and can maintain the social contacts that give meaning to their lives.

The third of the four disabilities is Williams's syndrome, a genetic defect with consequences less easily described but more profound than Usher's syndrome. People with Williams's syndrome have all their five senses but lack the ability to integrate their sensory universe into a quantitative framework. They do not live in the solid three-dimensional world that most of us take for granted. They have great difficulty in forming concepts of shape and size and number. They cannot draw pictures of things, and their world contains no mathematics. They have a characteristic facial appearance that marks them as different from other people. To compensate for these disabilities, many of them are verbally and musically gifted. They are also socially gifted. They have a childlike spontaneity and a cheerful temperament that enables them to make friends easily.

The fourth and most mysterious disability is autism. Autistic people have their senses unimpaired and have no difficulty with abstract concepts of shape and number. Their disability lies at a deeper level. They are born without the normal human ability to attach meaning to things that they see and hear and feel. They have great difficulty learning to talk, and many of them remain speechless all their lives. It happens that the leading character in the autism section of Oliver Sacks's program is Jessica Park, a lady whose family I have known since before she was born forty-five years ago. Her mother, Clara Park, has described her agonizingly slow development in two books, *The Siege* and *Exiting Nirvana* (1982, 2001), which are classics in the history of autism. Jessica was, as her mother wrote, "faced with a world in which an unreadable welter of impressions obscures even the distinction between objects and human beings." Like other autistic children who learn to speak, for many years Jessica used the pronouns "I" and "you" interchangeably. She had no concept of her own identity or of the identities of other people. Her mother recorded the fact that she used the word *heptagon* correctly before she used the word *yes.* Through the patient and devoted efforts of her parents and teachers, she has continued for forty-five years to learn new social skills and to increase her command of spoken language. Her intellectual growth has never stopped. Every year she becomes more independent and more capable of managing her own affairs. Through her paintings she is able to communicate glimpses of her inner world that cannot be communicated in words. Her paintings are exhibited and sold and bring her a modest income. In the television program we see her as she is today, after forty-five years of adapting to a world that is still largely beyond her comprehension. She speaks to a public audience and responds to questions. She is proud and happy because her gifts as an artist have been recognized. She talks about her paintings. Her speech sounds unnatural but is loud and clear.

What have these four disabilities to do with theology? Each of the four groups of people that I have described lives in a different world, and we normal people live in a fifth world different from theirs. Since we are the majority and have organized our world to suit our needs, they have been forced to adapt their ways of living so as to fit into our society as best they can. On the whole, they have adapted well to our world, but they still do not belong to it. They are aliens living here as guests. I find it illuminating to imagine the situations that would arise if the people with any one of the four disabilities were the majority and we were the minority. Wells already explored such a situation in his story "The Country of the Blind" a hundred years ago. If the majority were suffering either from achromatopsia or from Usher's syndrome, we would be in a situation similar to Wells's story. The gifts of color vision and hearing which seem to us so precious would have little value in the world of the achromatopes or the world of Usher's syndrome. In the Usher's world, our spoken language would be for our private use only. We would be forced to think in sign language in order to fit into the prevailing culture. But these first two disabilities are superficial compared with the third and fourth. The worlds of Williams's syndrome and autism differ from ours profoundly enough to require a different theology.

In the Williams world there is no mathematics and no science. Music and language flourish, but there is no concept of size or distance. The glories of the natural world are enjoyed but not analyzed. Nature is described in the language of art and poetry, not in the language of science. What kind of a theology can arise in the Williams world? We can imagine many possibilities for a Williams theology, all of them different from our own. Our Judeo-Christian theology begins with the first chapter of Genesis, with days that are numbered and counted. "And the evening and the morning were the first day," and the second day, and so on up to the seventh day. Our conceptions of God, like our conceptions of the universe, are rooted in an exact awareness of the passage of time. These conceptions are alien to the Williams world. A Williams theology would be more likely to resemble the theogony of the ancient Greeks, with gods riding in chariots across the sky and demigods hiding in bushes and caves on the earth.

In the autistic world there is no sin. Jessica Park's mother remarks on the fact that her word is absolutely trustworthy. Jessica cannot tell a deliberate lie because she has no concept of deceit. She cannot conceive of other people's thoughts and feelings, and so the idea of deceit cannot arise. There is no way for her to imagine doing deliberate harm to other people. When she hurts people, by losing her temper or throwing a tantrum, the hurt results from impatience and incomprehension, not from malice. If sin means deliberate malice, then Jessica is incapable of sin. When Jessica's father was asked whether she loves her family, he answered, "She loves us as much as she can." That is a precise statement of Jessica's condition. In the autistic world, humans love each other without understanding each other, and are incapable of hate. The theology of the autistic world must be radically different from Judeo-Christian theology. Since there is no sin, there can be no fall from grace and no redemption. Since other people's sufferings are unimaginable, the suffering of an incarnate God is also unimaginable. The autistic theology will probably be like Jessica's character, simple and transparent, concerned only with innocent joys and sorrows. The strongest link between Jessica's world and ours is that we share a common sense of humor and can laugh at each other's jokes.

The most important lesson for us to learn from imagining these alternative worlds is humility. In each of the four worlds, humans are well adapted to their situation and are totally unaware of what we consider to be their disabilities. They believe that they are well informed and aware of everything that is going on in the world around them. In the Williams and autistic worlds, I imagine them building religions and theologies to explain their world and their place in it. We know, of course, that they are unaware of huge and essential parts of their environment and are incapable of understanding what they cannot imagine. We know that their religions and theologies are deeply flawed because they are based on a partial view of reality. If we are honest, we must ask ourselves some hard questions. Why should we believe that our situation is different? Why should we believe that our view of reality is not equally partial, that our religion and theology are not equally flawed? How do we know that there are not huge and essential features of our universe and of our own nature of which we are equally unaware? Why should we believe that the processes of natural selection, which shaped us to survive the hazards of living in a world of fierce predators and harsh climates, should have given us a brain with a complete grasp of the universe we live in? These are the questions that neurology raises. Oliver Sacks has shown us glimpses of alien worlds. His glimpses are powerful arguments for the thesis that there may be more things in heaven and earth than we are capable of understanding. With this conclusion, William James and Sir John Templeton can both agree.

References

Baggaley, W. Jack. 2000. "Advanced Meteor Orbit Radar Observations of Interstellar Meteoroids." *Journal of Geophysical Research* 105:10353–61.

Berlin, Isaiah. 1953. *The Hedgehog and the Fox: An Essay on Tolstoy's View of History.* New York: Ivan Dee.

Blake, William. 1939. *Poetry and Prose of William Blake.* Ed. Geoffrey Keynes. London: Nonesuch Press. Quotations from pp. 118, 133, 182, 183.

Broecker, W. S. 1997. "Thermohaline Circulation, the Achilles Heel of Our Climate System: Will Man-Made CO_2 Upset the Climate Balance?" *Science* 278:1582-88.

Butler, Octavia E. 1995. *Parable of the Sower.* New York: Warner Books. Quotation from p. 22.

———. 1998. *Parable of the Talents.* New York: Seven Stories Press. Quotations from pp. 27, 249, 313.

Cahn, Robert W. 2005. "An Unusual Nobel Prize." *Notes and Records of the Royal Society* 59:145–53. Quotation from Kroemer on p. 150.

Chaisson, Eric. 1988. *Universe: An Evolutionary Approach to Cosmology.* Englewood Cliffs: Prentice Hall.

Dyson, Freeman J. 1979. "Time without End: Physics and Biology in an Open Universe." *Reviews of Modern Physics* 51:447–60.

———. 1999. *The Sun, the Genome, and the Internet: Tools of Scientific Revolutions.* New York: New York Public Library, Oxford University Press.

Frautschi, Steven. 1982. "Entropy in an Expanding Universe." *Science* 217:593-99.

Gold, Thomas. 1948. "Hearing, II. The Physical Basis of the Action of the Cochlea." *Proceedings of the Royal Society* 135:492–98.

Howard, A. W., et al. 2004. "Search for Nanosecond Optical Pulses from Nearby Solar-Type Stars." *Astrophysical Journal* 613:1270–84.

Hoyle, Fred. 1957. *The Black Cloud.* New York: Harper and Brothers.

Islam, Jamal N. 1977. "Possible Ultimate Fate of the Universe." *Quarterly Journal of the Royal Astronomical Society* 18:3.

———. 1983. *The Ultimate Fate of the Universe.* Cambridge: Cambridge University Press.

James, William. 1937. *The Varieties of Religious Experience: A Study in Human Nature; Being the Gifford Lectures on Natural Religion Delivered at Edinburgh in 1901–1902.* London and New York: Longmans Green. Quotation from p. 477.

Joy, Bill. 2000a. "Why the Future Doesn't Need Us." *Wired,* April 2000, 238–46.

———. 2000b. "Technology Check." *Washington Post,* April 18, 2000.

Kirschvink, Joseph L., et al. 2000. "A Low Temperature Transfer of ALH84001 from Mars to Earth." *Science* 290:791–95.

Krauss, Lawrence M., and Glenn D. Starkman. 1999a. "Life, the Universe, and Nothing: Life and Death in an Ever-Expanding Universe." Case Western Reserve University preprint CWRU–P1-99.

———. 1999b. "The Fate of Life in the Universe." *Scientific American* 281 (November): 58–65.

L'Engle, Madeleine. 1980. *Walking on Water: Reflections on Faith and Art.* Wheaton, Ill.: Harold Shaw. Quotations from pp. 65, 88-90, 106, 117–18, 129, 132, 148–49.

Lewis, C. S. 1952. *Out of the Silent Planet.* (Orig. pub. 1938.) London: Pan Books.

———. 1965. *Perelandra.* (Orig. pub. 1944.) New York: MacMillan.

———. 1972. *That Hideous Strength.* (Orig. pub. 1946.) New York: MacMillan. Quotations from pp. 7, 23.

Lhote, Henri. 1958. *À la découverte des fresques du Tassil.* Grenoble: Arthaud. Translated by Alan H. Brodrick as *The Search for the Tassili Frescoes: The Story of the Prehistoric Rock-Paintings of the Sahara.* New York: Dutton, 1959.

Manuel, Frank E. 1974. *The Religion of Isaac Newton.* Oxford: Clarendon Press. Quotation from pp. 135–36.

Margulis, Lynn. 1981. *Symbiosis in Cell Evolution.* San Francisco: Freeman and Co.

Margulis, Lynn, and Michael Dolan. 1997. "Swimming against the Current." *The Sciences* (New York Academy of Sciences) 37 (January): 20–25.

Meisel, David D., D. Janches, and J. D. Mathews. 2002. "Extrasolar Micrometeors Radiating from the Vicinity of the Local Interstellar Bubble." *Astrophysical Journal* 567:323–41.

Moravec, Hans. 1988. *Mind Children: The Future of Robot and Human Intelligence.* Cambridge, Mass.: Harvard University Press.

Munk, W. 2002. "Twentieth Century Sea Level: An Enigma." *Proceedings of the National Academy of Sciences* 99:6550–55.

Pagels, Elaine. 1979. *The Gnostic Gospels.* New York: Random House.

Park, Clara. 1982. *The Siege: The First Eight Years of an Autistic Child.* An Atlantic Monthly Press book. Boston: Little, Brown.

———. 2001. *Exiting Nirvana.* Boston: Little, Brown.

Pour-El, Marian B., and Ian Richards. 1981. "The Wave-Equation with Computable Initial Data Such That Its Unique Solution Is Not Computable." *Advances in Mathematics* 39:215–39.

Regis, Ed. 1990. *Great Mambo Chicken and the Transhuman Condition: Science Slightly Over the Edge.* Reading, Mass.: Addison-Wesley. Quotation from Moravec on p. 163.

Sacks, Oliver. 1995. *An Anthropologist on Mars: Seven Paradoxical Tales.* New York: Vintage Knopf.

———. 1998. *The Island of the Color-Blind.* (Orig. pub. 1996.) New York: Vintage Books, Random House. The television series was produced in six parts by the BBC in 1998 under the general title *The Mind Traveller.*

Schlesinger, W. H. 1977. "Carbon Balance in Terrestrial Detritus." *Annual Review of Ecology and Systematics* 8:51–81.

Schwartz, R. N., and C. H. Townes. 1961. "Interstellar and Interplanetary Communication by Optical Masers." *Nature* 190:205–8.

Scott, Henry P., et al. 2004. "Generation of Methane in the Earth's Mantle: In Situ High Pressure-Temperature Measurements of Carbonate Reduction." *Proceedings of the National Academy of Sciences* 101:14023–26.

Shih, W. M., J. D. Quispe, and G. F. Joyce. 2004. "A 1.7-kilobase Single-Stranded DNA That Folds into a Nanoscale Octahedron." *Nature* 427:618–21.

Silk, J., A. Szalay, and Y. Zeldovich. 1983. "The Large-Scale Structure of the Universe." *Scientific American* 249 (October): 72–80.

Stapeldon, Olaf. 1968. *Star Maker.* (Orig. pub. 1937.) New York: Dover, in one volume with *Last and First Men.*

———. 1972. *Sirius.* (Orig. pub. 1944.) New York: Dover, in one volume with *Odd John.*

Tsiolkovsky, Konstantin. 1895. *Dreams of Earth and Sky* (in Russian). Moscow: Goncharov. Ed. B. N. Vorobyeva. Moscow: USSR Academy of Sciences, 1959; quotations on pp. 40–41.

Von Neumann, John. 1948. "The General and Logical Theory of Automata." Unpublished lecture, in *Collected Works,* ed. A. H. Taub, 5:288–328. New York: Macmillan, 1961-63.

Woese, Carl R. 2004. "A New Biology for a New Century." *Microbiology and Molecular Biology Reviews* 68:173–86. Quotation about the child and the eddy on p. 176.

Index